I. M. Gelfand   A. Shen

Algebra

Birkhäuser
Boston · Basel · Berlin

Israel M. Gelfand
Department of Mathematics
Rutgers University
Piscataway, NJ 08854-8019
U.S.A.

Alexander Shen
Institute of Problems of
  Information Transmissions
Moscow
Russia

Library of Congress Cataloging-in-Publication Data
Gel'fand, I. M. (Izrail Moisevich)
    Algebra / I.M. Gelfand, Alexander Shen
       p. cm.
    ISBN 0-8176-3677-3 (acid-free) — ISBN 3-7643-3677-3 (acid-free)
    1. Algebra.   I. Shen, Alexander, 1958- . II. Title.
    QA152.2G45 1993   93-28904
    512–dc20   CIP

ISBN 0-8176-3677-3   Printed on acid-free paper.

Printed in the United States of America.   (MS/HP)

9 8 7 6   SPIN 11315704

Birkhäuser is part of *Springer Science+Business Media*
www.birkhauser.com

Acknowledgments

The authors wish to thank their friends and colleagues Mark Saul and Richard Askey whose experience as mathematicians and teachers were invaluable in the correction of this second printing.

Contents

Contents

Contents

1 Introduction

This book is about algebra. This is a very old science and its gems have lost their charm for us through everyday use. We have tried in this book to refresh them for you.

The main part of the book is made up of problems. The best way to deal with them is: Solve the problem by yourself – compare your solution with the solution in the book (if it exists) – go to the next problem. However, if you have difficulties solving a problem (and some of them are quite difficult), you may read the hint or start to read the solution. If there is no solution in the book for some problem, you may skip it (it is not heavily used in the sequel) and return to it later.

The book is divided into sections devoted to different topics. Some of them are very short, others are rather long.

Of course, you know arithmetic pretty well. However, we shall go through it once more, starting with easy things.

2 Exchange of terms in addition

Let's add 3 and 5:
$$3 + 5 = 8.$$

And now change the order:

$$5 + 3 = 8.$$

We get the same result. Adding three apples to five apples is the same as adding five apples to three – apples do not disappear and we get eight of them in both cases.

3 Exchange of terms in multiplication

Multiplication has a similar property. But let us first agree on notation. Usually in arithmetic, multiplication is denoted as "$\times$". In algebra this sign is usually replaced by a dot "$\cdot$". We follow this convention.

Let us compare $3 \cdot 5$ and $5 \cdot 3$. Both products are 15. But it is not so easy to explain why they are equal. To give each of three boys five apples is not the same as to give each of five boys three apples – the situations differ radically.

One of the authors of this book asked a seven-year-old girl, "How much is two times four?" "Eight", she answered immediately. "And four times two?" She started thinking, trying to add $2 + 2 + 2 + 2$. A year later she would know very well that the product remains the same when we exchange factors and she would forget that it was not so evident before.

The simplest way to explain why $5 \cdot 3 = 3 \cdot 5$ is to show a picture:

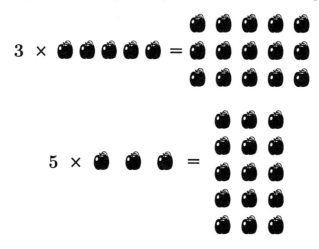

4   Addition in the decimal number system

If we want to know how much $7 + 9$ is, we may draw 7 apples and then 9 apples near them:

and then count all the apples together: one, two, three, four, ..., fifteen, sixteen. We get $7 + 9 = 16$. This method can be applied for any numbers; however, you need a lot of patience to try it on, say, 137 and 268. So mathematicians invented other methods. One of them is the standard addition method used in the positional number system.

4 Addition in the decimal number system

In different countries and at different times, people used different notations for numbers, and entire books are written about them. We are so used to the familiar decimal number system using the digits $0, 1, 2, \ldots, 8, 9$ that we don't realize how unbelievably convenient this convention has proved to be. Even the possibility of writing down very big numbers quickly was not self-evident for ancient people. A great mathematician of ancient Greece, Archimedes, even wrote a book called *The Sand Reckoner*. The main point of the book was to show that it is possible to write down the number that is greater than the number of sand particles filling the sphere whose radius is the distance between Earth and the stars.

Now the decimal number system has no rivals – except the binary number system, which is popular among computers, not people. This binary system has only two digits, 0 and 1 – but numbers have more digits. The computer does not worry about the length of numbers, but still wishes to keep rules of operation as simple as possible.

We shall speak about the binary system in another section, but now we return to our ordinary decimal system and to the addition method. We shall not explain it to you once more – you know it without us. Let us solve some problems instead.

Problem 1. Several digits "8" are written and some "+" signs are inserted to get the sum 1000. Figure out how it is done. (For example, if we try $88 + 88 + 8 + 8 + 88$, we fail because we get only 280 instead of 1000.)

Solution. Assume that

$$
\begin{array}{r}
\ldots 8 \\
\ldots \\
\ldots 8 \\
\hline
1000
\end{array}
$$

We do not know how many rows are here nor how many digits are used in each number. But we do know that each number ends with "8" and that the last digit of the sum is zero. How many numbers do we need to get this zero? If we use only one number, we get 8. If we use two numbers, we get 6 $(8 + 8 = 16)$, etc. To get zero we need at least five numbers:

$$
\begin{array}{r}
\ldots 8 \\
\ldots 8 \\
\ldots 8 \\
\ldots 8 \\
\underline{\ldots 8} \\
1000
\end{array}
$$

After we get this zero, we keep "4" in mind because $8+8+8+8+8 = 40$. To get the next zero in the "tens place" from this "4", we need to add at least two 8's since $4 + 8 + 8 = 20$.

$$
\begin{array}{r}
8 \\
8 \\
8 \\
..88 \\
\underline{..88} \\
1000
\end{array}
$$

We keep "2" in mind and we need only one more "8" to get 10:

$$
\begin{array}{r}
8 \\
8 \\
8 \\
88 \\
\underline{888} \\
1000
\end{array}
$$

The problem is solved: $8 + 8 + 8 + 88 + 888 = 1000$.

Problem 2. In the addition example

$$
\begin{array}{r}
A\,A\,A \\
\underline{B\,B\,B} \\
A\,A\,A\,C
\end{array}
$$

all A's denote some digit, all B's denote another digit and C denotes a third digit. What are these digits?

Solution. First of all A denotes 1 because no other digit can appear as a carry in the thousands position of the result. To find what B is let us ask ourselves: Do we get a (nonzero) carry adding the rightmost A and B? If we had no carry, we would get the same digit in the other two places (tens and hundreds), but this is not so. Therefore, the carry digit is not zero, and this is possible only if B = 9. Therefore we get the answer:

$$
\begin{array}{r}
111 \\
999 \\
\hline
1110
\end{array}
$$

5 The multiplication table and the multiplication algorithm

To compute the product of, say, 17 and 38, we may draw a picture of 17 rows, each containing 38 points, and then count all the points. But of course, nobody does this – we know an easier method of multiplying using the positional system.

This method (called the *multiplication algorithm*) is based on the multiplication table for digits and requires that you memorize the table. There is – sorry! – no way around it, and if, on being asked, "What is seven times eight?" in the middle of the night, you cannot answer "Fifty-six!" immediately, and instead try to add up seven eights half-asleep, we are unable to help you.

There is some good news, however. You don't need to memorize the product $17 \cdot 38$. Instead, you can compute it in two different ways:

$$
\begin{array}{r}
17 \\
38 \\
\hline
136 \\
51 \\
\hline
646
\end{array}
\qquad
\begin{array}{r}
38 \\
17 \\
\hline
266 \\
38 \\
\hline
646
\end{array}
$$

Both results are equal, though the intermediate results are different. A lucky coincidence, isn't it?

Here are some problems concerning multiplication.

Problem 3. A boy claims that he can multiply any three-digit number by 1001 instantly. If his classmate says to him "715" he gives the answer immediately. Compute this answer and explain the boy's secret.

Problem 4. Multiply 101010101 by 57.

Problem 5. Multiply 10001 by 1020304050.

Problem 6. Multiply 11111 by 1111.

Problem 7. A six-digit number having 1 as its leftmost digit becomes three times bigger if we take this digit off and put it at the end of the number. What is this number?

Solution. Look at the multiplication procedure:

$$
\begin{array}{r}
1\,A\,B\,C\,D\,E \\
3 \\
\hline
A\,B\,C\,D\,E\,1
\end{array}
$$

Here A,B,C,D and E denote some digits (we do not know whether all these digits are different or not). Digit E must be equal to 7, because among the products $3 \times 0 = 0$, $3 \times 1 = 3$, $3 \times 2 = 6$, $3 \times 3 = 9$, $3 \times 4 = 12$, $3 \times 5 = 15$, $3 \times 6 = 18$, $3 \times 7 = 21$, $3 \times 8 = 24$, $3 \times 9 = 27$ only $3 \times 7 = 21$ has the last digit 1. So we get:

$$
\begin{array}{r}
1\,A\,B\,C\,D\,7 \\
3 \\
\hline
A\,B\,C\,D\,7\,1
\end{array}
$$

When multiplying 7 by 3 we get a carry of 2, so $3 \times D$ must have its last digit equal to 5. This is possible only if $D = 5$:

$$
\begin{array}{r}
1\,A\,B\,C\,5\,7 \\
3 \\
\hline
A\,B\,C\,5\,7\,1
\end{array}
$$

In the same way, we find that $C = 8$, $B = 2$, $A = 4$. So we get the solution:

$$
\begin{array}{r}
1\ 4\ 2\ 8\ 5\ 7 \\
3 \\
\hline
4\ 2\ 8\ 5\ 7\ 1
\end{array}
$$

6 The division algorithm

Division is the most complicated thing among all the four arithmetic operations. To make yourself confident, you may try the following problems.

Problem 8. Divide 123123123 by 123. (Check your answer by multiplication!)

Problem 9. Can you predict the remainder when $111\ldots1$ (100 ones) is divided by 1111111?

Problem 10. Divide $1000\ldots0$ (20 zeros) by 7.

Problem 11. While solving the two preceding problems you may have discovered that quotient digits (and remainders) became periodic:

$$
\begin{array}{r}
\overset{\frown}{142857}\,14\ldots \\
7\,\overline{)\,100000000\ldots} \\
\underline{7} \\
30 \\
\underline{28} \\
20 \\
\underline{14} \\
60 \\
\underline{56} \\
40 \\
\underline{35} \\
50 \\
\underline{49} \\
10 \\
\underline{7} \\
30 \\
\underline{28} \\
2\ldots
\end{array}
$$

Is it just a coincidence, or will this pattern repeat?

Problem 12. Divide $2000\ldots000$ (20 zeros), $3000\ldots000$ (20 zeros), $4000\ldots000$ (20 zeros), etc. by 7. Compare the answers you get and explain what you see.

A multiplication fan may enjoy the following problem:

Problem 13. Multiply 142857 by 1, 2, 3, 4, 5, 6, 7, and look at the results. (It is easy to memorize these results and become a famous number cruncher who is able to multiply a random number, for example, 142857, by almost any digit!)

Problem 14. Try to invent similar tricks based on the division of $1000\ldots0$ by other numbers instead of 7.

7 The binary system

Problem 15. Find a generating rule, and write five or ten more lines:

$$0$$
$$1$$
$$10$$
$$11$$
$$100$$
$$101$$
$$110$$
$$111$$
$$1000$$
$$1001$$
$$1010$$
$$1011$$
$$1100$$
$$\ldots$$

Problem 16. You have weights of 1, 2, 4, 8, and 16 grams. Show that it is possible to get any weight from 0 to 31 grams using the following table ("+" means "the weight is used", "−" means "not used"):

	16	8	4	2	1	B	C
	<u>A</u>					<u>B</u>	<u>C</u>
0	−	−	−	−	−	00000	0
1	−	−	−	−	+	00001	1
2	−	−	−	+	−	00010	10
3	−	−	−	+	+	00011	11
4	−	−	+	−	−	00100	100
5	−	−	+	−	+	00101	101
6	−	−	+	+	−	00110	110
7	−	−	+	+	+	00111	111
8	−	+	−	−	−	01000	1000
9	−	+	−	−	+	01001	1001
10	−	+	−	+	−	01010	1010
11	−	+	−	+	+	01011	1011

$$\ldots$$

We can replace "−" by 0 and "+" by 1 (column B) and omit the leading zeros (column C). Then we get the same result as in the preceding problem.

7 The binary system

This table is called a conversion table between decimal and binary number systems:

Decimal	Binary
0	0
1	1
2	10
3	11
4	100
5	101
6	110
7	111
8	1000
9	1001
10	1010
11	1011
12	1100
...	...

Problem 17. What corresponds to 14 in the right column? What corresponds to 10000 in the left column?

The binary system has an advantage: you don't need to memorize as many as 10 digits; two is enough. But it has a disadvantage also: numbers are too long. (For example, 1024 is 10000000000 in binary.)

Problem 18. How is 45 (decimal) written in the binary system?

Problem 19. What (decimal) number is written as 10101101 in binary?

Problem 20. Try the usual addition method in binary version:

$$1010 \quad + \quad 101 = ?$$
$$1111 \quad + \quad 1 = ?$$
$$1011 \quad + \quad 1 = ?$$
$$1111 \quad + \quad 1111 = ?$$

Check your answers, converting all the numbers (the numbers being added and the sums) into the decimal system.

Problem 21. Try the usual subtraction algorithm in its binary version:

$$1101 \ - \ 101 = ?$$
$$110 \ - \ 1 = ?$$
$$1000 \ - \ 1 = ?$$

Check your answers, converting all the numbers into the decimal system.

Problem 22. Now try to multiply 1101 and 1010 (in binary):

$$\begin{array}{r} 1101 \\ 1010 \\ \hline ???? \end{array}$$

Check your result, converting the factors and the product into the decimal system.

Hint: Here are two patterns:

$$\begin{array}{r} 1011 \\ 11 \\ \hline 1011 \\ 1011 \\ \hline 100001 \end{array} \qquad \begin{array}{r} 1011 \\ 101 \\ \hline 1011 \\ 1011 \\ \hline 110111 \end{array}$$

Problem 23. Divide 11011 (binary) by 101 (binary) using the ordinary method. Check your result, converting all numbers into the decimal system.

Hint: Here is a pattern:

$$\begin{array}{r} 110 \quad \leftarrow \quad \text{the quotient} \\ 100\overline{)11001} \\ 100 \\ \hline 100 \\ 100 \\ \hline 1 \quad \leftarrow \quad \text{the remainder} \end{array}$$

Problem 24. In the decimal system the fraction 1/3 is written as 0.333.... What happens with 1/3 in the binary system?

8 The commutative law

Let us return to the rule "exchange of terms in addition does not change the sum". It can be written as

First term + Second term = Second term + First term

or in short

F.t. + S.t. = S.t. + F.t.

But even this short form seems too long for mathematicians, and they use single letters such as a or b instead of "F.t." and "S.t.". So we get

$$\boxed{a + b = b + a}$$

The law "exchange of factors does not change the product" can be written now as

$$\boxed{a \cdot b = b \cdot a}$$

Here "·" is a multiplication symbol. Often it is omitted:

$$ab = ba$$

The property $a + b = b + a$ is called the *commutative law for addition*; the property $ab = ba$ is called the *commutative law for multiplication*.

Remark. Sometimes it is impossible to omit the multiplication sign (·) in a formula; for example, $3 \cdot 7 = 21$ is not the same as $37 = 21$. By the way, multiplication had good luck in getting different symbols: the notations $a \times b$, $a \cdot b$, ab, and a*b (in computer programming) are all used.

9 The associative law

Now let us add three numbers instead of two:

$$3 + 5 + 11 = 8 + 11 = 19.$$

But there is another way:

$$3 + 5 + 11 = 3 + 16 = 19.$$

9 The associative law

Usually parentheses are used to show the desired order of operations:

$$(3 + 5) + 11$$

means that we have to add 3 and 5 first, and

$$3 + (5 + 11)$$

means that we have to add 5 and 11 first.

The result does not depend on the order of the operations. This fact is called the *associative law* by mathematicians. In symbols:

$$\boxed{(a + b) + c = a + (b + c)}$$

If you would like to have a real-life example, here it is. You can get sweet coffee with milk if you add milk to the coffee with sugar or if you add sugar to the coffee with milk. You get the same result – and this is the associative law:

$$(\text{sugar} + \text{coffee}) + \text{milk} = \text{sugar} + (\text{coffee} + \text{milk})$$

Problem 25. Try it.

Problem 26. Add $357 + 17999 + 1$ without paper and pencil.

Solution. It is not so easy to add 357 and 17999. But if you add $17999 + 1$, you get 18000 and now it is easy to add 357:

$$357 + (17999 + 1) = 357 + 18000 = 18357.$$

Problem 27. Add $357 + 17999$ without paper and pencil.

Solution. $357 + 17999 = (356 + 1) + 17999 = 356 + (1 + 17999) = 356 + 18000 = 18356$.

Problem 28. Add $899 + 1343 + 101$.

Hint. Remember the commutative law.

Multiplication is also associative:

$$(a \cdot b) \cdot c = a \cdot (b \cdot c)$$

or, in short,

$$(ab)c = a(bc).$$

Problem 29. Compute $37 \cdot 25 \cdot 4$.

Problem 30. Compute $125 \cdot 37 \cdot 8$.

10 The use of parentheses

A pedant is completely right saying that a notation like

$$2 \cdot 3 \cdot 4 \cdot 5$$

has no sense until we fix the order of operations. Even if we agree not to permute the factors, we have a lot of possibilities:

$$
\begin{aligned}
((2 \cdot 3) \cdot 4) \cdot 5 &= (6 \cdot 4) \cdot 5 = 24 \cdot 5 = 120 \\
(2 \cdot (3 \cdot 4)) \cdot 5 &= (2 \cdot 12) \cdot 5 = 24 \cdot 5 = 120 \\
(2 \cdot 3) \cdot (4 \cdot 5) &= 6 \cdot 20 = 120 \\
2 \cdot ((3 \cdot 4) \cdot 5) &= 2 \cdot (12 \cdot 5) = 2 \cdot 60 = 120 \\
2 \cdot (3 \cdot (4 \cdot 5)) &= 2 \cdot (3 \cdot 20) = 2 \cdot 60 = 120
\end{aligned}
$$

Problem 31. Find all possible ways to put parentheses in the product $2 \cdot 3 \cdot 4 \cdot 5 \cdot 6$ (not changing the order of factors; see the example just shown). Try to invent a systematic way of searching so as not to forget any possibilities.

Problem 32. How many "(" and ")" symbols do you need to specify completely the order of operations in the product

$$2 \cdot 3 \cdot 4 \cdot 5 \cdot 6 \cdots 99 \cdot 100 \,?$$

The parentheses are often omitted because the result is independent of the order of the operations. The reader may reconstruct them as he or she wishes.

The following problem shows what can be achieved by clever permutation and grouping.

Problem 33. Compute $1 + 2 + 3 + 4 + \cdots + 98 + 99 + 100$.

Solution. Group the 100 terms in 50 pairs: $1+2+3+4+\cdots+98+99+100 = (1+100) + (2+99) + (3+98) + \cdots + (49+52) + (50+51)$. Each pair has the sum 101. We have 50 pairs, so the total sum is $50 \cdot 101 = 5050$.

A legend says that as a schoolboy Karl Gauss (later a great German mathematician) shocked his school teacher by solving this problem instantly (as the teacher was planning to relax while the children were busy adding the hundred numbers).

11 The distributive law

There is one more law for addition and multiplication, called the *distributive law*. If two boys and three girls get 7 apples each, then the boys get $2 \cdot 7 = 14$ apples, the girls get $3 \cdot 7 = 21$ apples – and together they get

$$2 \cdot 7 + 3 \cdot 7 = 14 + 21 = 35$$

apples. The same answer can be computed in another way: each of $2 + 3 = 5$ children gets 7 apples, so the total number of apples is

$$(2 + 3) \cdot 7 = 5 \cdot 7 = 35 \,.$$

Therefore,

$$(2 + 3) \cdot 7 = 2 \cdot 7 + 3 \cdot 7$$

and, in general,

$$(a + b) \cdot c = a \cdot c + b \cdot c$$

This property is called the *distributive law*. Changing the order of factors we may also write

$$c \cdot (a + b) = c \cdot a + c \cdot b$$

Problem 34. Compute $1001 \cdot 20$ without pencil and paper.

Solution. $1001 \cdot 20 = (1000 + 1) \cdot 20 = 1000 \cdot 20 + 1 \cdot 20 = 20{,}000 + 20 = 20{,}020 \,.$

Problem 35. Compute $1001 \cdot 102$ without pencil and paper.

Solution. $1001 \cdot 102 = 1001 \cdot (100 + 2) = 1001 \cdot 100 + 1001 \cdot 2 = (1000 + 1) \cdot 100 + (1000 + 1) \cdot 2 = 100{,}000 + 100 + 2000 + 2 = 102{,}102 \,.$

The distributive law is a rule for removing brackets or parentheses. Let us see how it is used to transform the product of two sums

$$(a + b)(m + n) \,.$$

The number $(m + n)$ is the sum of the two numbers m and n and can replace c in the distributive law above:

$$(a + b) \cdot \boxed{c} = a \cdot \boxed{c} + b \cdot \boxed{c}$$

$$(a + b) \cdot \boxed{m + n} = a \cdot \boxed{m + n} + b \cdot \boxed{m + n}$$

Now we remember that $\boxed{m+n}$ is the sum of m and n and continue:

$$\ldots = a(m+n) + b(m+n) = am + an + bm + bn\,.$$

The general rule: To multiply two sums you need to multiply each term of the first sum by each term of the second one and then add all the products.

Problem 36. How many additive terms would be in

$$(a + b + c + d + e)(x + y + z)$$

after we use this rule?

12 Letters in algebra

In algebra we gradually make more and more use of letters (such as $a, b, c, \ldots, x, y, z$, etc.). Traditionally the use of letters (x's) is considered one of the most difficult topics in the school mathematics curriculum. Many years ago primary school pupils studied "arithmetic" (with no x's) and secondary school pupils started with "algebra" (with x's). Later "arithmetic" was renamed "mathematics" and x's were introduced (and created a mess, some people would say).

We hope that you, dear reader, never had difficulties understanding "what all these letters mean", but we still wish to give you some advice. If you ever want to explain the meaning of letters to your classmates, brothers and sisters, your parents, or your children (some day), just say that the letters are abbreviations for words. Let us explain what we mean.

In the equality

$$a + b = b + a$$

the letters a and b mean "the first term" and "the second term". When we write $a + b = b + a$ we mean that any numbers substituted instead of a and b give a true assertion. Therefore, $a + b = b + a$ can be considered as a unified short version of the equalities $1 + 7 = 7 + 1$ or $1028 + 17 = 17 + 1028$ as well as infinitely many other equalities of the same type.

Another example of the use of letters:

Problem 37. A small vessel and a big vessel contain (together) 5 liters. Two small and three big vessels contain together 13 liters. What are the volumes of the vessels?

Solution. (The "arithmetic" one.) The small and big vessels together contain 5 liters. Therefore, two small vessels and two big vessels together contain 10 liters ($10 = 2 \cdot 5$). As we know, two small vessels and three big vessels contain 13 liters. So we get 13 liters instead of 10 by adding one big vessel. Therefore the volume of a big vessel is 3 liters. Now it is easy to find the volume of a small vessel: together they contain 5 liters, so a small vessel contains $5 - 3 = 2$ liters. Answer: The volume of a small vessel is 2 liters, the volume of a big vessel is 3 liters.

This solution can be shortened if we use "Vol.SV" instead of "Volume of a Small Vessel" and "Vol.BV" instead of "Volume of a Big Vessel". Thus, according to the statement of the problem,

$$\text{Vol.SV} + \text{Vol.BV} = 5,$$

therefore

$$2 \cdot \text{Vol.SV} + 2 \cdot \text{Vol.BV} = 10.$$

We know also that

$$2 \cdot \text{Vol.SV} + 3 \cdot \text{Vol.BV} = 13.$$

If we subtract the preceding equality from the last one we find that $\text{Vol.BV} = 3$. Now the first equality implies that $\text{Vol.SV} = 5 - 3 = 2$.

Now the only thing to do is to replace our "Vol.SV" and "Vol.BV" by standard unknowns x and y – and we get the standard "algebraic" solution of our problem. Here it is: Denote the volume of a small vessel by x and the volume of a big vessel by y. We get the following system of equations:

$$\begin{aligned} x + y &= 5 \\ 2x + 3y &= 13. \end{aligned}$$

Multiplying the first equation by 2 we get

$$2x + 2y = 10$$

and subtracting the last equation from the second equation of our system we get

$$y = 13 - 10 = 3.$$

Now the first equation gives

$$x = 5 - y = 5 - 3 = 2.$$

Answer: $x = 2$, $y = 3$.

Finally, one more example of the use of letters in algebra.

"**Magic trick**". Choose any number you wish. Add 3 to it. Multiply the result by 2. Subtract the chosen number. Subtract 4. Subtract the chosen number once more. You get 2, don't you?

Problem 38. Explain why this trick is successful.

Solution. Let us follow what happens with the chosen number (we denote it by x):

Choose the number you wish	x
add 3 to it	$x + 3$
multiply the result by 2	$2 \cdot (x + 3) = 2x + 6$
subtract the chosen number	$(2x + 6) - x = x + 6$
subtract 4	$(x + 6) - 4 = x + 2$
subtract the chosen number once more. You get 2.	$(x + 2) - x = 2$

13 The addition of negative numbers

It is easy to check that $3 + 5 = 8$: just take three apples, add five apples, and count all the apples together: "one, two, three, four, ..., seven, eight". But how can we check that $(-3) + (-5) = (-8)$ or that $3 + (-5) = (-2)$? Usually this is explained by examples like the following two:

$3 + 5 = 8$	Yesterday it was $+3$. Today the temperature is 5 degrees warmer and is 8 degrees.
$(-3) + 5 = 2$	Yesterday it was -3 degrees. Today it is 5 degrees warmer, that is, $+2$.
$3 + (-5) = -2$	Yesterday was $+3$, today it is 5 degrees colder, that is, -2.
$(-3) + (-5) = (-8)$	Yesterday was -3, today it is 5 degrees colder, that is, -8.

(Here all temperatures are measured in Celsius degrees.)

Here is another example:

$3 + 5 = 8$	Three protons + five protons = = eight protons.
$(-3) + 5 = 2$	Three antiprotons + five protons = = two protons (ignoring γ-radiation).
$3 + (-5) = -2$	Three protons + five antiprotons = = two antiprotons (ignoring γ-radiation).
$(-3) + (-5) = (-8)$	Three antiprotons + five antiprotons = = eight antiprotons.

(Protons and antiprotons are elementary particles. When a proton meets an antiproton they annihilate one another, producing gamma radiation.)

14 The multiplication of negative numbers

To find how much three times five is, you add three numbers equal to five:

$$5 + 5 + 5 = 15.$$

The same explanation may be used for the product $1 \cdot 5$ if we agree that a sum having only one term is equal to this term. But it is evidently not applicable to the product $0 \cdot 5$ or $(-3) \cdot 5$: can you imagine a sum with zero or with minus three terms?

However, we may exchange the factors:

$$5 \cdot 0 = 0 + 0 + 0 + 0 + 0 = 0 \,,$$

$$5 \cdot (-3) = (-3) + (-3) + (-3) + (-3) + (-3) = -15 \,.$$

So if we want the product to be independent of the order of factors (as it was for positive numbers) we must agree that

$$0 \cdot 5 = 0, \quad (-3) \cdot 5 = -15 \,.$$

Now let us consider the product $(-3) \cdot (-5)$. Is it equal to -15 or to $+15$? Both answers may have advocates. From one point of view, even one negative factor makes the product negative – so if both factors are negative the product has a very strong reason to be negative. From the other point of view, in the table

$3 \cdot 5 = +15$	$3 \cdot (-5) = -15$
$(-3) \cdot 5 = -15$	$(-3) \cdot (-5) = ?$

we already have two minuses and only one plus; so the "equal opportunities" policy requires one more plus. So what?

Of course, these "arguments" are not convincing to you. School education says very definitely that minus times minus is plus. But imagine that your small brother or sister asks you, "Why?" (Is it a caprice of the teacher, a law adopted by Congress, or a theorem that can be proved?) You may try to answer this question using the following example:

$3 \cdot 5 = 15$	Getting five dollars three times is getting fifteen dollars.
$3 \cdot (-5) = -15$	Paying a five-dollar penalty three times is a fifteen-dollar penalty.
$(-3) \cdot 5 = -15$	Not getting five dollars three times is not getting fifteen dollars.
$(-3) \cdot (-5) = 15$	Not paying a five-dollar penalty three times is getting fifteen dollars.

14 The multiplication of negative numbers

Another explanation. Let us write the numbers

$$1, 2, 3, 4, 5, \ldots$$

and the same numbers multiplied by three:

$$3, 6, 9, 12, 15, \ldots$$

Each number is bigger than the preceding one by three. Let us write the same numbers in the reverse order (starting, for example, with 5 and 15):

5,	4,	3,	2,	1
15,	12,	9,	6,	3

Now let us continue both sequences:

5,	4,	3,	2,	1,	0,	−1,	−2,	−3,	−4,	−5, ...
15,	12,	9,	6,	3,	0,	−3,	−6,	−9,	−12,	−15, ...

Here −15 is under −5, so $3 \cdot (-5) = -15$; plus times minus is minus.

Now repeat the same procedure multiplying $1, 2, 3, 4, 5, \ldots$ by −3 (we know already that plus times minus is minus):

1,	2,	3,	4,	5
−3,	−6,	−9,	−12,	−15

Each number is three units less than the preceding one. Now write the same numbers in the reverse order:

5,	4,	3,	2,	1
−15,	−12,	−9,	−6,	−3

and continue:

5,	4,	3,	2,	1,	0,	−1,	−2,	−3,	−4,	−5, ...
−15,	−12,	−9,	−6,	−3,	0,	3,	6,	9,	12,	15, ...

Now 15 is under −5; therefore $(-3) \cdot (-5) = 15$.

Probably this argument would be convincing for your younger brother or sister. But you have the right to ask: So what? Is it possible to *prove* that $(-3) \cdot (-5) = 15$?

Let us tell the whole truth now. Yes, it is possible to prove that $(-3) \cdot (-5)$ *must be* 15 *if* we want the usual properties of addition, subtraction, and multiplication that are true for positive numbers to remain true for any integers (including negative ones).

Here is the outline of this proof: Let us prove first that $3 \cdot (-5) = -15$. What is -15? It is a number opposite to 15, that is, a number that produces zero when added to 15. So we must prove that

$$3 \cdot (-5) + 15 = 0.$$

Indeed,

$$3 \cdot (-5) + 15 = 3 \cdot (-5) + 3 \cdot 5 = 3 \cdot (-5 + 5) = 3 \cdot 0 = 0.$$

(When taking 3 out of the parentheses we use the law $ab + ac = a(b+c)$ for $a = 3$, $b = -5$, $c = 5$; we assume that it is true for all numbers, including negative ones.) So $3 \cdot (-5) = -15$. (The careful reader will ask why $3 \cdot 0 = 0$. To tell you the truth, this step of the proof is omitted – as well as the whole discussion of what zero is.)

Now we are ready to prove that $(-3) \cdot (-5) = 15$. Let us start with

$$(-3) + 3 = 0$$

and multiply both sides of this equality by -5:

$$((-3) + 3) \cdot (-5) = 0 \cdot (-5) = 0.$$

Now removing the parentheses in the left-hand side we get

$$(-3) \cdot (-5) + 3 \cdot (-5) = 0,$$

that is, $(-3) \cdot (-5) + (-15) = 0$. Therefore, the number $(-3) \cdot (-5)$ is opposite to -15, that is, is equal to 15. (This argument also has gaps. We should prove first that $0 \cdot (-5) = 0$ and that there is only one number opposite to -15.)

15 Dealing with fractions

If somebody asks you to compare the fractions

$$\frac{3}{5} \quad \text{and} \quad \frac{9}{15},$$

you would answer immediately that they are equal:

$$\frac{9}{15} = \frac{3 \cdot 3}{3 \cdot 5} = \frac{3}{5}.$$

But what would you say now: Are the fractions

$$\frac{221}{391} \quad \text{and} \quad \frac{403}{713}$$

equal or not?

If you remember the multiplication table for two-digit numbers, you would say immediately that they are equal:

$$\frac{221}{391} = \frac{17 \cdot 13}{17 \cdot 23} = \frac{13}{23} = \frac{31 \cdot 13}{31 \cdot 23} = \frac{403}{713}$$

But what are we to do if we do not remember this multiplication table? Then we should find the common denominator for the two fractions,

$$\frac{221}{391} = \frac{221 \cdot 713}{391 \cdot 713} \quad \text{and} \quad \frac{403}{713} = \frac{403 \cdot 391}{713 \cdot 391}$$

and compare numerators,

$$
\begin{array}{rr}
713 & 391 \\
221 & 403 \\
\hline
713 & 1173 \\
1426 & 1564 \\
1426 & \overline{157573} \\
\hline
157573 &
\end{array}
$$

After that we would know that the fractions are equal but would never discover that in fact they are equal to 13/23.

Problem 39. Which is bigger, 1/3 or 2/7?

Solution. $1/3 = 7/21$, $2/7 = 6/21$, so $1/3 > 2/7$.

The real-life version of this problem says, "Which is better, one bottle for three or two bottles for seven?" It suggests another solution: One bottle for three is equivalent to getting two bottles for six (and not for seven), so $1/3 > 2/7$. In scientific language, we found the "common numerator" instead of the common denominator:

$$\frac{1}{3} = \frac{2}{6} > \frac{2}{7}.$$

22

Problem 40. Which of the fractions

$$\frac{10001}{10002} \quad \text{and} \quad \frac{100001}{100002}$$

is bigger?

Hint. Both fractions are less than 1. What is the difference between them and 1?

Problem 41. Which of the fractions

$$\frac{12345}{54321} \quad \text{and} \quad \frac{12346}{54322}$$

is bigger?

Finding a common denominator is a traditional problem in teaching arithmetic. How much pie remains for you if your brother wants one-half and your sister wants one-third? The answer to this question is explained by the following picture:

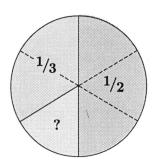

Generally speaking, you need to find a common denominator when adding fractions. It is a horrible error (which, of course, you avoid) to add numerators and denominators separately:

$$\frac{2}{3} + \frac{5}{7} \quad \longrightarrow \quad \frac{2+5}{3+7} = \frac{7}{10}.$$

Instead of the sum this operation gives you something in between the two fractions you started with ($7/10 = 0.7$ is between $2/3 = 0.666\ldots$ and $5/7 = 0.714285\ldots$).

This is easy to understand in a real-life situation. Assume that one team has two bottles for three people ($2/3$ for each) and the other team has five bottles for seven people (5/7 for each). After they meet they have something in between ($2 + 5$ bottles for $3 + 7$ people).

Problem 42. Fractions $\dfrac{a}{b}$ and $\dfrac{c}{d}$ are called neighbor fractions if their difference $\dfrac{ad - bc}{bd}$ has numerator ± 1, that is, $ad - bc = \pm 1$. Prove that

(a) in this case neither fraction can be simplified (that is, neither has any common factors in numerator and denominator);

(b) if $\dfrac{a}{b}$ and $\dfrac{c}{d}$ are neighbor fractions, then $\dfrac{a + b}{c + d}$ is between them and is a neighbor fraction for both $\dfrac{a}{b}$ and $\dfrac{c}{d}$; moreover,

(c) no fraction $\dfrac{e}{f}$ with positive integer e and f such that $f < b + d$ is between $\dfrac{a}{b}$ and $\dfrac{c}{d}$.

Problem 43. A stick is divided by red marks into 7 equal segments and by green marks into 13 equal segments. Then it is cut into 20 equal pieces. Prove that any piece (except the two end pieces) contains exactly one mark (which may be red or green).

Solution. End pieces carry no marks because $\dfrac{1}{20}$ is smaller than $\dfrac{1}{7}$ and $\dfrac{1}{13}$. We have 18 other pieces – and it remains to prove that none of them can have more than one mark. (We have 18 marks – 6 red and 12 green – so no piece will be left without a mark.) Red marks correspond to numbers of the form $\dfrac{k}{7}$, green marks correspond to numbers of the form $\dfrac{l}{13}$. A fraction

$$\frac{k + l}{7 + 13} = \frac{k + l}{20}$$

is between them and is a cut point dividing these marks. Therefore, two marks of different colors cannot belong to the same piece. Two marks of the same color also cannot appear on one piece because the distance between them (either 1/7 or 1/13) is bigger than the piece length 1/20.)

Problem 44. What is better, to get five percent of seven billion or seven percent of five billion?

Problem 45. How can you cut from a 2/3-meter-long string a piece of length 1/2 meter, without having a meter stick?

Solution. A piece of length 1/2 m constitutes three-fourths of the whole string:

$$\frac{3}{4}\cdot\frac{2}{3}=\frac{2}{4}=\frac{1}{2}$$

and you need to cut off one-fourth of the string.

16 Powers

In the sequence of numbers

$$2,\ 4,\ 8,\ 16,\ldots$$

each number is twice as large as the preceding one:

$$
\begin{aligned}
4 &= 2\cdot 2\\
8 &= 4\cdot 2 = 2\cdot 2\cdot 2 \ (3\text{ factors})\\
16 &= 8\cdot 2 = 2\cdot 2\cdot 2\cdot 2 \ (4\text{ factors})\\
&\ \ldots
\end{aligned}
$$

Mathematicians use the following useful notation:

$$
\begin{aligned}
2\cdot 2 &= 2^2\\
2\cdot 2\cdot 2 &= 2^3\\
2\cdot 2\cdot 2\cdot 2 &= 2^4\\
&\ \ldots
\end{aligned}
$$

so, for example, $2^6 = 64$.

Now the sequence $2,4,8,16,\ldots$ can be written as $2,2^2,2^3,2^4,\ldots$. We read a^n as "a to the n-th power" or "the n-th power of a"; a is called the *base*, and n is called an *exponent*.

There are special names for a^2 and a^3. They are "a squared" and "a cubed", respectively. (A square with side a has area a^2; a cube with edge a has volume a^3.)

Problem 46. Compute: (a) 2^{10}; (b) 10^3; (c) 10^7.

Problem 47. How many decimal digits do you need to write down 10^{1000}?

Astronomers use powers of 10 to write big numbers in a short form. For example, the speed of light is about 300,000 kilometers per second $= 3 \cdot 10^5$ km/s $= 3 \cdot 10^8$ m/s $= 3 \cdot 10^{10}$ cm/s.

Problem 48. In astronomy the distance covered by light in one year is called a *light-year*. What is the distance (approximately) between the Sun and the closest star measured in meters if it is about 4 light-years?

17 Big numbers around us

The number of molecules in one gram of water	$\simeq 3 \cdot 10^{22}$
The radius of Earth	$\simeq 6 \cdot 10^6$ m
The distance between Earth and the Moon	$\simeq 4 \cdot 10^8$ m
The distance between Earth and the Sun (the "astronomical unit")	$\simeq 1.5 \cdot 10^{11}$ m
The radius of the part of the universe observed up to now	$\simeq 10^{26}$ m
The mass of Earth	$\simeq 6 \cdot 10^{24}$ kg
The age of Earth	$\simeq 5 \cdot 10^9$ years
The age of the universe	$\simeq 1.5 \cdot 10^{10}$ years
The number of people on Earth	$\simeq 5 \cdot 10^9$
The average duration of a human life	$\simeq 2 \cdot 10^9$ seconds

Remark. When speaking of big numbers you must keep in mind that the same quantity may be big or small, depending on the unit you choose. For example, the distance between Earth and the Sun, measured in light-years, is about 0.000015 lt-yr, or, in meters (as seen from the table above), $1.5 \cdot 10^{11}$ m.

We shall see later that not only big numbers but also small numbers can be written conveniently using powers.

Programmers prefer to deal with powers of 2 (and not of 10). It turns out that $2^{10} = 1024$ is rather close to $1000 = 10^3$. So the prefix *kilo*, which usually means 1000 (1 kilogram = 1000 grams, 1 kilometer

= 1000 meters, etc.), means "1024" in programming: 1 kilobyte is 1024 bytes.

Problem 49. (a) How many decimal digits do you need to write down 2^{20}? (b) How many for the number 2^{100}? (c) Draw the graph showing how the number of decimal digits in 2^n depends on n.

(To answer the last question, the number of decimal digits in 2^n is approximately $0.3\,n$: $2^{10} \simeq 10^{0.3 \cdot 10}$, $2^n \simeq 10^{0.3\,n}$. Remember this when studying logarithms.)

Many types of pocket calculators use powers of 10 to show the product of two big numbers. For example,

$$370{,}000 \cdot 2{,}100{,}000 = 7.77 \cdot 10^{11},$$

but on the screen you do not see the dot and the base 10, just

$$7.77 \qquad 11 \qquad \text{or} \qquad 7.77 \qquad \text{E}11$$

because of the screen limitations. In the usual form

$$777000000000$$

the answer would overflow the calculator screen.

18 Negative powers

We have seen the sequence of powers of 2:

$$2,\ 4,\ 8,\ 16,\ 32,\ 64,\ 128,\ \ldots$$

Now let us start with some number of the sequence (for example, 128) and write it in the reverse order:

$$128,\ 64,\ 32,\ 16,\ 8,\ 4,\ 2.$$

In the first sequence each number was two times bigger than the preceding one; in the second each number is two times smaller than the preceding one. Let us continue this sequence:

$$128,\ 64,\ 32,\ 16,\ 8,\ 4,\ 2,\ 1,\ \frac{1}{2},\ \frac{1}{4},\ \frac{1}{8},\ \frac{1}{16},\ \ldots$$

The sequence

$$2, 4, 8, 16, 32, 64, 128, \ldots$$

could be written as

$$2, 2^2, 2^3, 2^4, 2^5, 2^6, 2^7, \ldots.$$

In reverse order,
$$128, 64, 32, 16, 8, 4, 2$$

could be written as

$$2^7, 2^6, 2^5, 2^4, 2^3, 2^2, 2.$$

The analogy suggests the following continuation:

$$128, \quad 64, \quad 32, \quad 16, \quad 8, \quad 4, \quad 2, \quad 1, \quad \tfrac{1}{2}, \quad \tfrac{1}{4}, \quad \tfrac{1}{8}, \quad \tfrac{1}{16}, \quad \cdots$$
$$2^7, \quad 2^6, \quad 2^5, \quad 2^4, \quad 2^3, \quad 2^2, \quad 2^1. \quad 2^0, \quad 2^{-1}, \quad 2^{-2}, \quad 2^{-3}, \quad 2^{-4}, \ldots$$

This notation is widely used. So, for example,

$$2^3 = 8, \quad 2^1 = 2, \quad 2^0 = 1, \quad 2^{-1} = \frac{1}{2}, \quad 2^{-2} = \frac{1}{4}, \quad 2^{-3} = \frac{1}{8}, \text{ etc.}$$

When we spoke about powers before we said that 2^3 is "2 used 3 times as a factor" and 2^5 is "2 used 5 times as a factor". We can even say that 2^1 is "2 used once as a factor", but for 2^0 or 2^{-1} such an explanation cannot be taken seriously. It is just an agreement between mathematicians to understand 2^{-n} (for positive integer n) as $\dfrac{1}{2^n}$.

We hope that this agreement seems rather natural to you. Later we shall see that it is convenient and – in a sense – unavoidable.

Problem 50. Write down (a) 10^{-1}; (b) 10^{-2}; (c) 10^{-3} as decimal fractions.

19 Small numbers around us

1 cm	$= 10^{-2}\,\text{m}$
1 mm	$= 10^{-3}\,\text{m}$
1 μm	$= 10^{-6}\,\text{m}$
1 nanometer	$= 10^{-9}\,\text{m}$
1 angstrom	$= 10^{-10}\,\text{m}$
The mass of a water molecule	$\simeq 3 \cdot 10^{-23}\,\text{g}$
The size of a living cell	$\simeq 15$ to $350 \cdot 10^{-9}\,\text{m}$
The size at which modern physical laws become inapplicable (the "elementary length", as physicists say)	$\simeq 10^{-31}\,\text{cm}$
The wavelength of red light	$\simeq 7 \cdot 10^{-7}\,\text{m}$

As we have said already, there is no difference, in principle, between "big" and "small" numbers. For example, Earth's radius is about $6 \cdot 10^3$ km and at the same time about $4 \cdot 10^{-5}$ astronomical units.

Now let us return to the general definition of powers.

Definition. For positive integers n,

$$a^n = a \cdot a \cdots a \quad (n \text{ times})$$
$$a^{-n} = \frac{1}{a^n}$$
$$a^0 = 1$$

Problem 51. Is the equality $a^{-n} = \frac{1}{a^n}$ valid for negative n and for $n = 0$?

Is it possible to *prove* that $a^{-n} = \frac{1}{a^n}$? No, because the notation a^{-n} makes no sense without an agreement (called a *definition* by mathematicians). If suddenly all mathematicians change their mind and agree to understand a^{-n} in another way, then the equality $a^{-n} = \frac{1}{a^n}$ would be false. But you may be sure that this would never happen because nobody wants to violate such a convenient agreement. We would get into a mess if we did so.

Our notation allows us to write the long expression

$$2 \cdot a \cdot a \cdot a \cdot a \cdot b \cdot b \cdot b \cdot c \cdot c \cdot d$$

in the shorter form

$$2a^4b^3c^2d$$

and also rewrite

$$\frac{2 \cdot a \cdot a \cdot a \cdot a \cdot c \cdot c}{b \cdot b \cdot b \cdot d}$$

as

$$2a^4b^{-3}c^2d^{-1}.$$

Problem 52. Write the short form for the following expressions:

(a) $\quad a \cdot a \cdot a \cdot a \cdot a \cdot a \cdot a \cdot a \cdot a \cdot b \cdot b \cdot b \cdot b$

(b) $\quad \dfrac{2 \cdot a \cdot a \cdot a}{b \cdot b}$

Answer: (a) $a^{10}b^4$; (b) $2a^3b^{-2}$.

Problem 53. Rewrite using only positive powers:

(a) a^3b^{-5}; (b) $a^{-2}b^{-7}$.

Answer: (a) $\dfrac{a^3}{b^5}$; (b) $\dfrac{1}{a^2b^7}$.

20 How to multiply a^m by a^n, or why our definition is convenient

It is easy to multiply a^m by a^n if m and n are positive. For example,

$$a^5 \cdot a^3 = \underbrace{(a \cdot a \cdot a \cdot a \cdot a)}_{5 \text{ times}} \cdot \underbrace{(a \cdot a \cdot a)}_{3 \text{ times}} = a^8.$$

In general, $a^m \cdot a^n = a^{m+n}$ (indeed, a^m is a repeated m times and a^n is a repeated n times). Also

$$a^m \cdot a^1 = a^m \cdot a = a^{m+1}.$$

But the powers may also be negative. It turns out that our rule is valid in this case, too. For example, for $m = 5$, $n = -3$, it states that

$$a^5 \cdot a^{-3} = a^{5+(-3)} = a^2.$$

Let us check it: By definition, $a^5 \cdot a^{-3}$ is

$$a^5 \cdot \frac{1}{a^3} = \frac{a \cdot a \cdot a \cdot a \cdot a}{a \cdot a \cdot a} = a^2.$$

More pedantic readers would ask us to check also that

$$a^{-5} \cdot a^3 = a^{-5+3} = a^{-2}.$$

O.K. By definition,

$$a^{-5} \cdot a^3 = \frac{1}{a^5} \cdot a^3 = \frac{a^3}{a^5} = \frac{1}{a^2} = a^{-2}.$$

Even more pedantic readers would remember that both numbers m and n may be negative and ask to check, for example, that

$$a^{-5} \cdot a^{-3} = a^{(-5)+(-3)} = a^{-8}.$$

Indeed,

$$a^{-5} \cdot a^{-3} = \frac{1}{a^5} \cdot \frac{1}{a^3} = \frac{1}{a^8} = a^{-8}.$$

Don't relax – there are still other cases. One of the exponents (or even both) may be equal to zero, and a^0 was defined by a special agreement. So let us check that

$$a^m \cdot a^0 = a^{m+0} = a^m.$$

Indeed, $a^0 = 1$ by definition, so

$$a^m \cdot a^0 = a^m \cdot 1 = a^m.$$

Question. Is it necessary to consider the cases $m < 0$, $m = 0$ and $m > 0$ in the last argument separately?

Problem 54. Find a formula for $\dfrac{a^m}{a^n}$. Is your answer valid for all integers m and n?

21 The rule of multiplication for powers

When multiplying powers with the same base, you need to add exponents:

$$a^m \cdot a^n = a^{m+n}$$

This rule can be used to multiply small and big numbers in a convenient way. For example, to multiply $2 \cdot 10^7$ and $3 \cdot 10^{-11}$ we multiply 2 and 3 and add 7 and -11:

$$(2 \cdot 10^7) \cdot (3 \cdot 10^{-11}) = (2 \cdot 3) \cdot (10^7 \cdot 10^{-11}) = 6 \cdot 10^{7+(-11)} = 6 \cdot 10^{-4}.$$

This method is used in computers (but with base 2 instead of 10).

Problem 55. (a) You know that $2^{1001} \cdot 2^n = 2^{2000}$. What is n?

(b) You know that $2^{1001} \cdot 2^n = 1/4$. What is n?

(c) Which is bigger: 10^{-3} or 2^{-10}?

(d) You know that $\dfrac{2^{1000}}{2^n} = 2^{501}$. What is n?

(e) You know that $\dfrac{2^{1000}}{2^n} = 1/16$. What is n?

(f) You know that $4^{100} = 2^n$. What is n?

(g) You know that $2^{100} \cdot 3^{100} = a^{100}$. What is a?

(h) You know that $\left(2^{10}\right)^{15} = 2^n$. What is n?

We said earlier that the definition of negative powers is in a sense unavoidable. Now we shall explain what we mean. Assume that we want to define negative power in some way, but want to keep the rule $a^{m+n} = a^m \cdot a^n$ true for all m and n. It turns out that the only way to do so is to follow our definition. Indeed, for $n = 0$ we must have $a^m \cdot a^0 = a^{m+0}$, that is, $a^m \cdot a^0 = a^m$. Therefore, $a^0 = 1$. But then $a^n \cdot a^{-n} = a^{n+(-n)} = a^0 = 1$ implies that $a^{-n} = 1/a^n$.

What do we get if the power is used as a base for another power? For example,

$$(a^2)^3 = \underbrace{a^2 \cdot a^2 \cdot a^2}_{3 \text{ times}} = (a \cdot a) \cdot (a \cdot a) \cdot (a \cdot a) = a^6.$$

Similarly,

$$(a^m)^n = a^{m \cdot n}$$

for any positive m, n. And again our conventions "think for us": the same formula is also true for negative m and n. For example,

$$(a^{-2})^3 = \left(\frac{1}{a^2}\right)^3 = \frac{1}{a^2} \cdot \frac{1}{a^2} \cdot \frac{1}{a^2} = \frac{1}{a^6} = a^{-6} = a^{(-2)\cdot 3}.$$

Problem 56. Check this formula for other combinations of signs (if $m > 0$, $n < 0$; if both m and n are negative; if one of them is equal to zero).

The last formula about powers:

$$\boxed{(ab)^n = a^n \cdot b^n}$$

Problem 57. Check this formula for positive and negative integers n.

Problem 58. What is $(-a)^{775}$? Is it a^{775} or $-a^{775}$?

Problem 59. Invent a formula for $\left(\dfrac{a}{b}\right)^n$.

Now a^n is defined for any integer n (positive or not) and for any a. But that is not the end of the game, because n may be a number that is not an integer.

Problem 60. Give some suggestions: What might $4^{1/2}$ be? And $27^{1/3}$? Motivate your suggestions as well as you can.

The definition of $a^{m/n}$ will be given later. (But that also is not the last possible step.)

22 Formula for short multiplication: The square of a sum

As we have seen already,

$$(a + b)(m + n) = am + an + bm + bn$$

(to multiply two sums you must multiply each term of the first sum by each term of the second sum and then add all the products). Now consider the case when the letters inside the parentheses are the same:

$$(a + b)(a + b) = aa + ab + ba + bb.$$

Remember that $ab = ba$ and that aa and bb are usually denoted as a^2 and b^2; we get
$$(a + b)(a + b) = a^2 + 2ab + b^2,$$
or
$$\boxed{(a + b)^2 = a^2 + 2ab + b^2}$$

Problem 61. (a) Compute 101^2 without pencil and paper.

(b) Compute 1002^2 without pencil and paper.

Problem 62. Each of the two factors of a product becomes 10 percent bigger. How does the product change?

The rule in words: "The square of the sum of two terms is the sum of their squares plus two times the product of the terms".

Be careful here: "the square of the sum" and "the sum of the squares" sound very similar, but are different; the square of the sum is $(a + b)^2$ and the sum of the squares is $a^2 + b^2$.

Problem 63. Are the father of the son of NN and the son of the father of NN the same person?

23 How to explain the square of the sum formula to your younger brother or sister

A kind wizard liked to talk with children and to make them gifts. He was especially kind when many children came together; each of them got as many candies as the number of children. (So if you came alone, you got one, and if you came with a friend you got two and your friend got two.)

Once, a boys came together. Each of them got a candies – together they got a^2 candies. After they went away with the candies, b girls came and got b candies each – so the girls together got b^2 candies. So that day, the boys and girls got $a^2 + b^2$ candies together.

The next day, a boys and b girls decided to come together. Each of $a+b$ children got $a+b$ candies, so all the children together got $(a+b)^2$ candies. Did they get more or fewer candies than yesterday – and how big is the difference?

To answer this question we may use the following argument. The second time, each of the a boys got b more candies (because of the b girls), so all the boys together got ab more candies. Each girl got a more candies (because of the a boys), so all the girls got ba additional candies. So together, the boys and girls got $ab + ba = 2ab$ candies more than on the previous day. So $(a + b)^2$ is $2ab$ more than $a^2 + b^2$, that is, $(a + b)^2 = a^2 + b^2 + 2ab$.

Problem 64. Cut a square with edge $a + b$ into one square $a \times a$, one square $b \times b$ and two rectangles $a \times b$.

Solution.

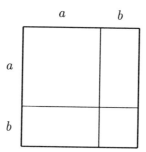

The formula $(a+b)^2 = a^2 + b^2 + 2ab$ may be considered as a generic formula for infinitely many equalities like $(5+7)^2 = 5^2 + 2 \cdot 5 \cdot 7 + 7^2$ or $(13 + \frac{1}{3})^2 = 13^2 + 2 \cdot 13 \cdot \frac{1}{3} + \left(\frac{1}{3}\right)^2$; we get these equalities by replacing a and b by specific numbers. These number may, of course, be negative. For example, for $a = 7$, $b = -5$ we get

$$(7 + (-5))^2 = 7^2 + 2 \cdot 7 \cdot (-5) + (-5)^2.$$

Plus times minus is minus, and minus times minus is plus, so we get

$$(7 - 5)^2 = 7^2 - 2 \cdot 7 \cdot 5 + 5^2.$$

The same thing could be done for any other numbers, so the general rule is that

$$\boxed{(a - b)^2 = a^2 - 2ab + b^2}$$

Or in words: "The square of the difference is the sum of the squares minus two times the product of the terms".

Problem 65. Compute (a) 99^2; (b) 998^2 without pencil and paper.

Problem 66. What do the formulas $(a + b)^2 = a^2 + 2ab + b^2$ and $(a - b)^2 = a^2 - 2ab + b^2$ give when (a) $a = b$; (b) $a = 2b$?

24 The difference of squares

Problem 67. Multiply $a + b$ and $a - b$.

Solution. $(a + b)(a - b) = a(a - b) + b(a - b) = a^2 - ab + ba - b^2 = a^2 - b^2$ (here ab and ba compensate for each other). So we get the formula

$$\boxed{a^2 - b^2 = (a + b)(a - b)}$$

Problem 68. Multiply $101 \cdot 99$ without pencil and paper.

Problem 69. A piece of size $b \times b$ was cut from an $a \times a$ square.

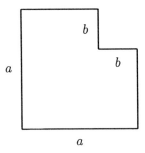

Cut the remaining part into pieces and combine the pieces into a rectangle with sides $a - b$ and $a + b$.

These three formulas – the square of a sum, the square of a difference, and the difference of squares – are called "short multiplication formulas".

Problem 70. Two integers differ by 2. Multiply them and add 1 to the product. Prove that the result is a perfect square (the square of an integer). For example,

$$3 \cdot 5 + 1 = 16 = 4^2,$$
$$13 \cdot 15 + 1 = 196 = 14^2.$$

Solution. (First version.) Let n denote the smaller number. Then the other number is $n + 2$. Their product is $n(n + 2) = n^2 + 2n$. Adding 1, we get $n^2 + 2n + 1 = (n + 1)^2$ (the formula for the square of the sum).

(Second version.) Let n denote the bigger number. Then the smaller one is $n - 2$. The product is $n(n - 2) = n^2 - 2n$. Adding 1 we get $n^2 - 2n + 1 = (n - 1)^2$ (the square of the difference formula).

(Third version.) If we want to be fair and not choose between the bigger and the smaller number, let us denote by n the number halfway between the numbers. Then the smaller number is $n - 1$, the bigger one is $n + 1$, and the product is $(n + 1)(n - 1) = n^2 - 1$ (the difference of squares formula), that is, it is a perfect square minus one.

Problem 71. Write the sequence of squares of $1, 2, 3, \ldots$:

$$1, \ 4, \ 9, \ 16, \ 25, \ 36, \ 49, \ \ldots$$

and under any two consecutive numbers of this sequence write their difference:

$$
\begin{array}{ccccccc}
1 & 4 & 9 & 16 & 25 & 36 & 49 \ \ldots \\
 & 3 & 5 & 7 & 9 & 11 & 13 \ \ldots
\end{array}
$$

In the second sequence any two consecutive numbers differ by 2. Can you explain why?

Solution. The consecutive numbers n and $n + 1$ have squares n^2 and $(n + 1)^2 = n^2 + 2n + 1$. The difference between these squares is $2n + 1$, and it becomes greater by 2 if we add 1 to n.

Remark. A sequence where each term is greater than the preceding one by a fixed constant (as in $3, 5, 7, 9, \ldots$) is called an arithmetic (pronounced "arithmEtic", not "arIthmetic") progression. We shall meet progressions again later.

Problem 72. There is a rule that allows us to square any number with the last digit 5, namely, "Drop this last digit out and get some n; multiply n by $n + 1$ and add the digits 2 and 5 to the end". For example, for 35^2, we delete 5 and get 3, multiplying 3 and 4 we get 12, adding "2" and "5" we get the answer: 1225. Explain why this rule works.

Problem 73. Compute $(a + b + c)^2$.

Solution. $(a + b + c)^2 = (a + b + c)(a + b + c) = a^2 + ab + ac + ba + b^2 + bc + ca + cb + c^2 = a^2 + b^2 + c^2 + 2ab + 2ac + 2bc$.

	a	b	c
a	a^2	ab	ac
b	ba	b^2	bc
c	ca	cb	c^2

Problem 74. Compute $(a + b - c)^2$.

Hint. Use the answer of the preceding problem.

Problem 75. Compute $(a + b + c)(a + b - c)$.

Hint. Use the difference-of-squares formula.

Problem 76. Compute $(a + b + c)(a - b - c)$.

Hint. The difference-of-squares formula is useful here also.

Problem 77. Compute $(a + b - c)(a - b + c)$.

Hint. Even here the difference-of-squares formula can be used!

Problem 78. Compute $(a^2 - 2ab + b^2)(a^2 + 2ab + b^2)$.

Solution. This is equal to

$$(a - b)^2(a + b)^2 = ((a - b)(a + b))^2 = (a^2 - b^2)^2 = a^4 - 2a^2b^2 + b^4.$$

Another solution:

$$(a^2 - 2ab + b^2)(a^2 + 2ab + b^2) =$$
$$= ((a^2 + b^2) + 2ab)((a^2 + b^2) - 2ab) = (a^2 + b^2)^2 - (2ab)^2 =$$
$$= a^4 + 2a^2b^2 + b^4 - 4a^2b^2 = a^4 + b^4 - 2a^2b^2.$$

25 The cube of the sum formula

Let us derive the formula for $(a + b)^3$. By definition,

$$(a + b)^3 = (a + b)(a + b)(a + b),$$

and we may start here. But part of the job is done already:

$$(a + b)^3 = (a + b)^2(a + b) = (a^2 + 2ab + b^2)(a + b).$$

Now we have to multiply each term of the first sum by each term of the second one and take the sum of all products:

$$(a^2 + 2ab + b^2)(a + b) =$$
$$= a^2 \cdot a + 2ab \cdot a + b^2 \cdot a +$$
$$+ a^2 \cdot b + 2ab \cdot b + b^2 \cdot b.$$

Remembering how to multiply powers with a common base (that is, that $a^m \cdot a^n = a^{m+n}$) and putting a-factors first, we get

$$a^3 + 2a^2b + ab^2 +$$
$$+ a^2b + 2ab^2 + b^3.$$

Here some terms are similar (only the numerical factors are different); they are written one under another. Collecting them, we get

$$\boxed{(a + b)^3 = a^3 + 3a^2b + 3ab^2 + b^3}$$

Problem 79. Compute 11^3 without pencil and paper.

Hint. $11 = 10 + 1$.

Problem 80. Compute 101^3 without pencil and paper.

Problem 81. Compute $(a - b)^3$.

Solution. We may compute it in the same way as before, writing $(a - b)^3 = (a - b)^2(a - b) = (a^2 - 2ab + b^2)(a - b)$ etc. But an easier way is to substitute $(-b)$ for b in the formula for $(a + b)^3$:

$$(a + (-b))^3 = a^3 + 3a^2 \cdot (-b) + 3a(-b)^2 + (-b)^3$$

or

$$\boxed{(a - b)^3 = a^3 - 3a^2b + 3ab^2 - b^3}$$

(recall that minus times minus is plus and plus times minus is minus).

26 The formula for $(a+b)^4$

Before computing $(a+b)^4$ let us try to guess the answer. To do so, look at the formulas we already have:

$$(a+b)^2 = a^2 + 2ab + b^2$$
$$(a+b)^3 = a^3 + 3a^2b + 3ab^2 + b^3.$$

To get more "experimental data" we can add the formula

$$(a+b)^1 = a + b.$$

So we have:

$$(a+b)^1 = a + b$$
$$(a+b)^2 = a^2 + 2ab + b^2$$
$$(a+b)^3 = a^3 + 3a^2b + 3ab^2 + b^3$$
$$(a+b)^4 = ???$$

How many additive terms do you expect in $(a+b)^4$? Five, of course. What is the first term? Definitely, a^4. The next term is a more difficult puzzle. (To tell you the truth, it will be $4a^3b$.) To explain how it can be guessed let us divide our question into two parts:

(1) What powers of a and b will appear?

(2) What numeric coefficients will appear?

Part (1) is simpler. If the formula for

$(a+b)^1$ uses a and b,
$(a+b)^2$ uses a^2, ab and b^2,
$(a+b)^3$ uses a^3, a^2b, ab^2 and b^3,

we may expect that

$(a+b)^4$ uses a^4, a^3b, a^2b^2, ab^3, and b^4.

Now look at the coefficients (we write the factor "1" to make our formulas more uniform):

$$(a+b)^1 = 1a + 1b$$
$$(a+b)^2 = 1a^2 + 2ab + 1b^2$$
$$(a+b)^3 = 1a^3 + 3a^2b + 3ab^2 + 1b^3$$

or, without terms (only the coefficients):

$$
\begin{array}{ccccc}
1 & 1 & & & \\
1 & 2 & 1 & & \\
1 & 3 & 3 & 1 & \\
? & ? & ? & ? & ?
\end{array}
$$

(we have already said that we expect five terms in the $(a+b)^4$ formula). The first coefficient is, of course, 1. It seems that the second is 4 (because in the second column we have 1, 2 and 3). So we get

$$
\begin{array}{ccccc}
1 & 1 & & & \\
1 & 2 & 1 & & \\
1 & 3 & 3 & 1 & \\
1 & 4 & ? & ? & ?
\end{array}
$$

Two more coefficients can be guessed. In $(a+b)^4$, the letters a and b have equal rights, so b^4 must have the same coefficient as a^4, and ab^3 must have the same coefficient as a^3b – to avoid discrimination:

$$
\begin{array}{ccccc}
1 & 1 & & & \\
1 & 2 & 1 & & \\
1 & 3 & 3 & 1 & \\
1 & 4 & ? & 4 & 1.
\end{array}
$$

Now only a^2b^2 remains, and if we cannot guess it, we must compute it by brute force:

$$(a+b)^4 = (a+b)^3(a+b) = (a^3 + 3a^2b + 3ab^2 + b^3)(a+b) =$$

$$
\begin{aligned}
&= a^3 \cdot a & + \ 3a^2b \cdot a & \quad + \ 3ab^2 \cdot a & \quad + \ b^3 \cdot a & \quad + \\
& & + \ a^3 \cdot b & \quad + \ 3a^2b \cdot b & \quad + \ 3ab^2 \cdot b & \quad + \ b^3 \cdot b = \\
&= a^4 & + \ 3a^3b & \quad + \ 3a^2b^2 & \quad + \ ab^3 & \quad + \\
& & + \ a^3b & \quad + \ 3a^2b^2 & \quad + \ 3ab^3 & \quad + \ b^4
\end{aligned}
$$

(again the similar terms are written one under another). Collecting them, we get

$$\boxed{(a+b)^4 = a^4 + 4a^3b + 6a^2b^2 + 4ab^3 + b^4}$$

All our guesses turn out to be true and we find the remaining coefficient of a^2b^2, which turns out to be 6.

27 Formulas for $(a+b)^5$, $(a+b)^6$, ... and Pascal's triangle

In $(a+b)^5$ we expect terms

$$a^5 \quad a^4b \quad a^3b^2 \quad a^2b^3 \quad ab^4 \quad b^5$$

with coefficients

$$1 \quad 5 \quad ? \quad ? \quad 5 \quad 1$$

To find the two remaining coefficients (they are expected to be equal, of course) let us proceed as usual:

$$(a+b)^5 = (a^4 + 4a^3b + 6a^2b^2 + 4ab^3 + b^4)(a+b) =$$

$$= a^4 \cdot a + 4a^3b \cdot a + 6a^2b^2 \cdot a + 4ab^3 \cdot a + b^4 \cdot a +$$
$$+ a^4 \cdot b + 4a^3 \cdot b + 6a^2b^2 \cdot b + 4ab^3 \cdot b + b^4 \cdot b =$$

$$= a^5 + 5a^4b + 10a^3b^2 + 10a^2b^3 + 5ab^4 + b^5.$$

So our table of coefficients has one more row:

$$
\begin{array}{cccccc}
1 & 1 & & & & \\
1 & 2 & 1 & & & \\
1 & 3 & 3 & 1 & & \\
1 & 4 & 6 & 4 & 1 & \\
1 & 5 & 10 & 10 & 5 & 1 \\
\end{array}
$$

Probably you have already figured out the rule: Each coefficient is equal to the sum of the coefficient above it and the one to the left of it: $1+4=5$, $4+6=10$, $6+4=10$, $4+1=5$.

The reason this is so becomes clear if we look at our computation ignoring everything except coefficients:

$$1\dots + 4\dots + 6\dots + 4\dots + 1\dots +$$
$$+ 1\dots + 4\dots + 6\dots + 4\dots + 1\dots =$$

$$1\dots + 5\dots + 10\dots + 10\dots + 5\dots + 1\dots$$

They are added exactly as the rule says.

For aesthetic reasons, we may write the table in a more symmetric way and add "1" on the top (because $(a+b)^0 = 1$). We get a triangle

$$
\begin{array}{ccccccccccc}
 & & & & & 1 & & & & & \\
 & & & & 1 & & 1 & & & & \\
 & & & 1 & & 2 & & 1 & & & \\
 & & 1 & & 3 & & 3 & & 1 & & \\
 & 1 & & 4 & & 6 & & 4 & & 1 & \\
1 & & 5 & & 10 & & 10 & & 5 & & 1
\end{array}
$$

which can be continued using the rule that each number is the sum of the two numbers immediately above it (except for the first and the last numbers, which are equal to 1). For example, the next row will be

$$
\begin{array}{ccccccc}
1 & 6 & 15 & 20 & 15 & 6 & 1
\end{array}
$$

and it corresponds to the formula

$$\boxed{(a+b)^6 = a^6 + 6a^5b + 15a^4b^2 + 20a^3b^3 + 15a^2b^4 + 6ab^5 + b^6}$$

This triangle is called *Pascal's triangle* (Blaise Pascal [1623–1662] was a French mathematician and philosopher.)

Problem 82. Compute 11^3, 11^4, 11^5 and 11^6.

Problem 83. Write a formula for $(a+b)^7$.

Problem 84. Find formulas for $(a-b)^4$, $(a-b)^5$ and $(a-b)^6$.

Problem 85. Compute the sums of all the numbers in the first, second, third, etc., rows of Pascal's triangle. Can you see the rule? Can you explain the rule?

Problem 86. What do the formulas for $(a+b)^2$, $(a+b)^3$, $(a+b)^4$, etc., give when $a = b$?

Problem 87. Do you see the connection between the two preceding problems?

Problem 88. What do the formulas for $(a+b)^2$, $(a+b)^3$, $(a+b)^4$, etc., give when $a = -b$?

28 Polynomials

By a *polynomial* we mean an expression containing letters (called *variables*), numbers, addition, subtraction and multiplication. Here are some examples:

$$a^4 + a^3b + ab^3 + b^4$$

$$(5 - 7x)(x - 1)(x - 3) + 11$$

$$(a + b)(a^3 + b^3)$$

$$(a + b)(a + 2b) + ab$$

$$(x + y)(x - y) + (y - x)(y + x)$$

$$0$$

$$(x + y)^{100}$$

These examples contain not only addition, subtraction and multiplication, but also positive integer constants as powers. These are legal because they can be considered as shortcuts (for example, a^4 may be considered as short notation for $a \cdot a \cdot a \cdot a$, which is perfectly legal). But a^{-7} or x^y are *not* polynomials.

A *monomial* is a polynomial that does not use addition or subtraction, that is, a product of letters and numbers. Here are some examples of monomials:

$$5 \cdot a \cdot 7 \cdot b \cdot a$$

$$127a^{15}$$

$$(-2)a^2b$$

(in the last example the minus sign is not subtraction but a part of the number "-2").

Usually numbers and identical letters are collected: for example, $5 \cdot a \cdot 7 \cdot b \cdot a$ is written as $35a^2b$.

Please keep in mind that a monomial is a polynomial, so sometimes for a mathematician one ("mono") is many ("poly").

Each polynomial can be converted into the sum of monomials if we remove parentheses. For example,

$$(a + b)(a^3 + b^3) = aa^3 + ab^3 + ba^3 + bb^3 = a^4 + ab^3 + ba^3 + b^4,$$

$$(a + b)(a + 2b) = a^2 + 2ab + ba + 2b^2.$$

When doing so we can get similar monomials (having the same letters with the same powers and differing only in the coefficients). For example, in the second polynomial above, the terms $2ab$ and ba are similar. They can be collected into $3ab$ and we get

$$(a + b)(a + 2b) = a^2 + 2ab + ba + 2b^2 = a^2 + 3ab + 2b^2.$$

Problem 89. Convert $(1 + x - y)(12 - zx - y)$ into a sum of monomials and collect the similar terms.

Solution.

$$(1 + x - y)(12 - zx - y) =$$
$$= 12 - zx \underline{-y} + 12x - xzx - xy \underline{-12y} + yzx + y^2 =$$
$$= 12 - zx - 13y + 12x - zx^2 - yx + yzx + y^2.$$

(The similar terms are underlined.)

Strictly speaking, this is not enough, because we need a sum of monomials and now we have subtraction. Therefore we need to do one more step to get

$$12 + (-1)zx + (-13)y + 12x + (-1)zx^2 + (-1)yx + 1yzx + 1y^2$$

(to make the terms more uniform we added the factor "1" before xyz and before y^2).

A *standard form* of a polynomial is a sum of monomials, where each monomial is a product of a number (called a *coefficient*) and of powers of different letters, and where all similar monomials are collected.

To add two polynomials in standard form we must add the coefficients of similar terms. If we get a zero coefficient, the corresponding term vanishes:

$$(1x + (-1)y) + (1y + (-2)x + 1z) =$$
$$(1 + (-2))x + ((-1) + 1)y + 1z = (-1)x + 0y + 1z = (-1)x + 1z.$$

To multiply two polynomials in standard form we need to multiply each term of the first polynomial by each term of the second polynomial. When multiplying monomials, we add powers of each variable:

$$(a^5 b^7 c) \cdot (a^3 bd^4) = a^{5+3} b^{7+1} cd^4 = a^8 b^8 cd^4.$$

After this is done, we have to collect similar terms. For example,

$$(x - y)(x^2 + xy + y^2) = x^3 + \underline{x^2y} + \underline{\underline{xy^2}} - \underline{yx^2} - \underline{\underline{xy^2}} - y^3 = x^3 - y^3.$$

(The pedantic reader may find that we have violated the rules adopted for the standard form of a polynomial, because the coefficients -1 and 1 are omitted.)

Problem 90.

(a) Multiply $(1 + x)(1 + x^2)$.

(b) Multiply $(1 + x)(1 + x^2)(1 + x^4)(1 + x^8)$.

(c) Compute $(1 + x + x^2 + x^3)^2$.

(d) Compute $(1 + x + x^2 + x^3 + \cdots + x^9 + x^{10})^2$.

(e) Find the coefficients of x^{30} and x^{29} in
$(1 + x + x^2 + x^3 + \cdots + x^9 + x^{10})^3$.

(f) Multiply $(1 - x)(1 + x + x^2 + x^3 + \cdots + x^9 + x^{10})$.

(g) Multiply $(a + b)(a^2 - ab + b^2)$.

(h) Multiply $(1 - x + x^2 - x^3 + x^4 - x^5 + x^6 - x^7 + x^8 - x^9 + x^{10})$
by $(1 + x + x^2 + x^3 + x^4 + x^5 + x^6 + x^7 + x^8 + x^9 + x^{10})$.

29 A digression: When are polynomials equal?

The word "equal" for polynomials may be understood in many different ways. The first possibility: Polynomials are equal if they can be transformed into one another by using algebraic rules (removing parentheses, collecting similar terms, finding common factors, and so on). Another possibility: Two polynomials are considered to be equal if after substituting any numbers for the variables they have the same numeric value. It turns out that these two definitions are equivalent; if two polynomials are equal in the sense of one of these definitions they are also equal in the sense of the other one. Indeed, if one polynomial can be converted into another using algebraic transformations, these transformations are still valid when variables are replaced by numbers.

So these polynomials have the same numeric value after replacement. It is not easy to prove the reverse statement: *If two polynomials are equal for any values of variables, they can be converted into each other by algebraic transformations.* So we shall use it – sorry! – without proof.

If we want to convince somebody that two given polynomials are equal, the first version of the definition is preferable; it is enough to show the sequence of algebraic transformations needed to get the second polynomial from the first one. On the other hand, if we want to convince somebody that two polynomials are not equal, the second definition is better; it is enough to find numbers that lead to the different values of the polynomials.

Problem 91. Prove that

$$(x - 1)(x - 2)(x - 3)(x - 4) \neq (x + 1)(x + 2)(x + 3)(x + 4)$$

without computations.

Solution. When $x = 1$ the left-hand side is equal to zero and the right-hand side is not, therefore these polynomials are not equal according to the second definition.

Problem 92. In the (true) equality

$$(x^2 - 1)(x + \cdots) = (x - 1)(x + 3)(x + \cdots)$$

some numbers are replaced by dots. What are these numbers?

Hint. Substitute -1 and -3 for x.

Now assume that somebody gives us two polynomials, not saying whether they are different or equal. How can we check this? We can try to substitute different numbers for the variables. If at least once these polynomials have different numeric values we can be sure that they are different. Otherwise we may suspect that these polynomials are in fact equal.

Problem 93. George tries to check whether the polynomials $(x + 1)^2 - (x - 1)^2$ and $x^2 + 4x - 1$ are equal or not by substituting 1 and -1 for x. Is it a good idea?

Solution. No. These polynomials have equal values for $x = -1$ (both values are -4) and for $x = 1$ (both give 4). However, they are not equal; for example, they have different values for $x = 0$.

To check whether two polynomials are equal or not in a more regular way, we may convert them to a standard form. If after this they differ only in the order of the monomials (or in the order of the factors inside the monomials), then the polynomials are equal. If not, it is possible to prove that the polynomials are different.

Sometimes equal polynomials are called "identically equal", meaning that they are equal for all values of variables. So, for example, $a^2 - b^2$ is identically equal to $(a - b)(a + b)$.

Remark. Later we shall see that sometimes a finite number of tests is enough to decide whether two polynomials are equal or not.

30 How many monomials do we get?

Problem 94. Each of two polynomials contains four monomials. What is the maximal possible number of monomials in their product?

Remark. Of course, any polynomial can be extended by monomials with zero coefficients like this:

$$x^3 + 4 = x^3 + 0x^2 + 0x + 4$$

Such monomials are ignored.

Solution. Multiply $(a + b + c + d)$ by $(x + y + z + u)$:
$$(a + b + c + d)(x + y + z + u) =$$
$$= ax + ay + az + au +$$
$$bx + by + bz + bu +$$
$$cx + cy + cz + cu +$$
$$dx + dy + dz + du.$$

We get 16 terms. It is clear that 16 is the maximum possible number (because each of 4 monomials of the first polynomial is multiplied by each of 4 monomials of the second one).

Problem 95. Each of two polynomials contains four monomials. Is it possible that their product contains fewer than 16 monomials?

Solution. Yes, if there are similar monomials among the products. For example,

$$(1 + x + x^2 + x^3)(1 + x + x^2 + x^3) = 1 + 2x + 3x^2 + 4x^3 + 3x^4 + 2x^5 + x^6,$$

that is, after collecting similar terms we get 7 monomials instead of 16.

Problem 96. Is it possible when multiplying two polynomials that, after collecting similar terms, all terms vanish (have zero coefficients)?

Answer. No.

Remark. Probably this problem seems silly; it is clear that it cannot happen. If you think so, please reconsider the problem several years from now.

Problem 97. Is it possible when multiplying two polynomials that after the collection of similar terms all terms vanish (have zero coefficients) except one? (Do not count the case when each of the polynomials has only one monomial.)

Problem 98. Is it possible that the product of two polynomials contains fewer monomials than each of the factors?

Solution. Yes:

$$
\begin{aligned}
(x^2 + 2xy + 2y^2)(x^2 - 2xy + 2y^2) &= \\
&= ((x^2 + 2y^2) + 2xy)((x^2 + 2y^2) - 2xy) = \\
&= (x^2 + 2y^2)^2 - (2xy)^2 = \\
&= x^4 + 4x^2y^2 + 4y^4 - 4x^2y^2 = \\
&= x^4 + 4y^4.
\end{aligned}
$$

31 Coefficients and values

Recall Pascal's triangle and the formulas for $(a + b)^n$ for different n:

$$
\begin{array}{llll}
1 & (a+b)^0 & = & 1 \\
1\ 1 & (a+b)^1 & = & 1a + 1b \\
1\ 2\ 1 & (a+b)^2 & = & 1a^2 + 2ab + 1b^2 \\
1\ 3\ 3\ 1 & (a+b)^3 & = & 1a^3 + 3a^2b + 3ab^2 + 1b^3 \\
1\ 4\ 6\ 4\ 1 & (a+b)^4 & = & 1a^4 + 4a^3b + 6a^2b^2 + 4ab^3 + 1b^4
\end{array}
$$

etc. Each of these formulas is an equality between two polynomials.

Problem 99. What do we get for $a = 1$, $b = 1$?

Solution.

$$
\begin{aligned}
(1+1)^0 &= 1 \\
(1+1)^1 &= 1+1 \\
(1+1)^2 &= 1+2+1 \\
(1+1)^3 &= 1+3+3+1 \\
(1+1)^4 &= 1+4+6+4+1
\end{aligned}
$$

etc. Recall that $1 + 1 = 2$; so we proved that the sum of any row of Pascal's triangle is a power of 2. For example, the sum of the 25th row of Pascal's triangle is equal to 2^{24}.

Problem 100. Add the numbers of some row of Pascal's triangle with alternating signs. You get 0:

$$
\begin{aligned}
1 - 1 &= 0 \\
1 - 2 + 1 &= 0 \\
1 - 3 + 3 - 1 &= 0 \\
1 - 4 + 6 - 4 + 1 &= 0
\end{aligned}
$$

etc. Why does this happen?

Hint. Try $a = 1$, $b = -1$.

Problem 101. Imagine that the polynomial $(1 + 2x)^{200}$ is converted to the standard form (the sum of powers of x with numerical coefficients). What is the sum of all the coefficients?

Hint. Try $x = 1$.

Problem 102. The same question for the polynomial $(1 - 2x)^{200}$ instead of $(1 + 2x)^{200}$.

Problem 103. Imagine that the polynomial $(1 + x - y)^3$ is converted to the standard form. What is the sum of its coefficients?

Problem 104. (*continued*) What is the sum of the coefficients of the terms not containing y?

Problem 105. (*continued*) What is the sum of the coefficients of the terms containing x?

32 Factoring

To multiply polynomials you may need a lot of patience, but you do not need to think; just follow the rules carefully. But to reconstruct factors if you know only their product you sometimes need a lot of ingenuity. And some polynomials cannot be decomposed into a product of nontrivial (nonconstant) factors at all. The decomposition process is called *factoring*, and there are many tricks that may help. We'll show some tricks now.

Problem 106. Factor the polynomial $ac + ad + bc + bd$.

Solution. $ac + ad + bc + bd = a(c+d) + b(c+d) = (a+b)(c+d)$.

Problem 107. Factor the following polynomials:
(a) $ac + bc - ad - bd$;
(b) $1 + a + a^2 + a^3$;
(c) $1 + a + a^2 + a^3 + \cdots + a^{13} + a^{14}$;
(d) $x^4 - x^3 + 2x - 2$.

Sometimes we first need to cut one term into two pieces before it is possible to proceed.

Problem 108. Factor $a^2 + 3ab + 2b^2$.

Solution. $a^2 + 3ab + 2b^2 = a^2 + ab + 2ab + 2b^2 = a(a+b) + 2b(a+b)$
$= (a + 2b)(a + b)$.

Remark. When multiplying two polynomials we collect the similar terms into one term. So it is natural to expect that when going in the other direction we may have to split a term into a sum of several terms.

Problem 109. Factor:
(a) $a^2 - 3ab + 2b^2$;
(b) $a^2 + 3a + 2$.

The formula for the square of the sum can be read "from right to left" as an example of factoring: the polynomial $a^2 + 2ab + b^2$ is factored into $(a+b)(a+b)$. The same factorization can also be obtained as follows:

$$a^2 + 2ab + b^2 = a^2 + ab + ab + b^2 = a(a + b) + b(a + b) = (a + b)(a + b).$$

Problem 110. Factor:

(a) $a^2 + 4ab + 4b^2$;

(b) $a^4 + 2a^2b^2 + b^4$;

(c) $a^2 - 2a + 1$.

Sometimes it is necessary to add and subtract some monomial (reconstructing the annihilated terms). We show this trick factoring $a^2 - b^2$ (though we know the factorization in advance: it is the difference-of-squares formula):

$$a^2 - b^2 = a^2 - ab + ab - b^2 = a(a-b) + b(a-b) = (a+b)(a-b).$$

Problem 111. Factor $x^5 + x + 1$.

Solution. $x^5 + x + 1 = x^5 + x^4 + x^3 - x^4 - x^3 - x^2 + x^2 + x + 1 = x^3(x^2+x+1) - x^2(x^2+x+1) + (x^2+x+1) = (x^3 - x^2 + 1)(x^2+x+1)$.

Probably you are discouraged by this solution because it seems impossible to invent it. The authors share your feeling.

Let us look at the factorization $a^2 - b^2 = (a+b)(a-b)$ once more from another viewpoint. If $a = b$, then the right-hand side is equal to zero (one of the factors is zero). Therefore the left-hand side must be zero, too. Indeed, $a^2 = b^2$ when $a = b$. Similarly, if $a + b = 0$ then $a^2 = b^2$ (in this case $a = -b$ and $a^2 = b^2$ because in changing the sign we do not change the square).

Problem 112. Prove that if $a^2 = b^2$ then $a = b$ or $a = -b$.

The moral of this story: When trying to factor a polynomial it is wise to see when it has a zero value. This may give you an idea what the factors might be.

Problem 113. Factor $a^3 - b^3$.

Solution. The expression $a^3 - b^3$ has a zero value when $a = b$. So it is reasonable to expect a factor $a - b$. Let us try: $a^3 - b^3 = a^3 - a^2b + a^2b - ab^2 + ab^2 - b^3 = a^2(a-b) + ab(a-b) + b^2(a-b) = (a^2 + ab + b^2)(a-b)$.

Problem 114. Factor $a^3 + b^3$.

Solution. $a^3 + b^3 = a^3 + a^2b - a^2b - ab^2 + ab^2 + b^3 = a^2(a+b) - ab(a+b) + b^2(a+b) = (a^2 - ab + b^2)(a+b)$.

The same factorization can be obtained from the solution of the preceding problem by substituting $(-b)$ for b.

Problem 115. Factor $a^4 - b^4$.

Solution. $a^4 - b^4 = a^4 - a^3b + a^3b - a^2b^2 + a^2b^2 - ab^3 + ab^3 - b^4 = a^3(a-b) + a^2b(a-b) + ab^2(a-b) + b^3(a-b) = (a-b)(a^3 + a^2b + ab^2 + b^3)$.

Problem 116. Factor:

(a) $a^5 - b^5$;

(b) $a^{10} - b^{10}$;

(c) $a^7 - 1$.

Another factorization of $a^4 - b^4$:

$$a^4 - b^4 = (a^2 - b^2)(a^2 + b^2).$$

These two factorizations are in fact related; both can be obtained from

$$(a^4 - b^4) = (a - b)(a + b)(a^2 + b^2)$$

by a suitable grouping of factors.

Problem 117. Factor $a^2 - 4b^2$.

Solution. Using that $4 = 2^2$ we write:

$$a^2 - 4b^2 = a^2 - 2^2b^2 = a^2 - (2b)^2 = (a - 2b)(a + 2b)$$

Let us try to apply the same trick to $a^2 - 2b^2$. Here we need a number called "the square root of two" and denoted by $\sqrt{2}$. It is approximately equal to $1.4142\ldots$; its main property is that its square is equal to 2: $(\sqrt{2})^2 = 2$. (Generally speaking, a square root of a nonnegative number a is defined as a nonnegative number whose square is equal to a. It is denoted by $\sqrt{a}$. Such a number always exists and is defined uniquely; see below.)

Using the square root of two we may write:

$$a^2 - 2b^2 = a^2 - (\sqrt{2}b)^2 = (a - \sqrt{2}b)(a + \sqrt{2}b).$$

So we are able to factor $a^2 - 2b^2$, though we are forced to use $\sqrt{2}$ as a coefficient.

Remark. Look at the equality

$$a - b = (\sqrt{a})^2 - (\sqrt{b})^2 = (\sqrt{a} - \sqrt{b})(\sqrt{a} + \sqrt{b}).$$

So we have factored $a-b$, haven't we? No, we haven't, because $\sqrt{a}-\sqrt{b}$ is not a polynomial; taking the square root is not a legal operation for polynomials – only addition, subtraction and multiplication are allowed. But how about $a-\sqrt{2}b$? Why do we consider it as a polynomial? Because our definition of a polynomial allows it to be constructed from letters and numbers using addition, subtraction, and multiplication. And $\sqrt{2}$ is a perfectly legal number (though it is defined as a square root of another number). So in this case everything is O.K.

Problem 118. Factor: (a) a^2-2; (b) a^2-3b^2; (c) $a^2+2ab+b^2-c^2$; (d) $a^2 + 4ab + 3b^2$.

Problem 119. Factor $a^4 + b^4$. (The known factorization of $a^4 - b^4$ seems useless because substituting $(-b)$ for b we get nothing new.)

Solution. A trick: add and subtract $2a^2b^2$. It helps:

$$a^4 + b^4 = a^4 + 2a^2b^2 + b^4 - 2a^2b^2 =$$
$$= (a^2 + b^2)^2 - (\sqrt{2}ab)^2 = (a^2 + b^2 + \sqrt{2}ab)(a^2 + b^2 - \sqrt{2}ab).$$

Let us see what we now know. We can factor $a^n - b^n$ for any positive integer n (one of the factors is $a - b$). If n is odd, the substitution of $-b$ for b gives a factorization of $a^n + b^n$ (one of the factors is $a + b$). But what about $a^2 + b^2$, $a^4 + b^4$, $a^6 + b^6$, etc.? We have just factored the second one.

Problem 120. Can you factor any other polynomial of the form $a^{2n} + b^{2n}$?

Hint. $a^6 + b^6 = (a^2)^3 + (b^2)^3$. The same trick may be used if n has an odd divisor greater than 1.

But the simplest case, $a^2 + b^2$, remains unsolved. It would be possible to write

$$a^2 + b^2 = a^2 - (\sqrt{-1} \cdot b)^2 = (a - \sqrt{-1} \cdot b)(a + \sqrt{-1} \cdot b)$$

if a square root of -1 exists. But – alas – it is not the case (the square of any nonzero number is positive and therefore not equal to -1). But mathematicians are tricky; if such a number does not exist, it must be invented. So they invented it, and got new numbers called *complex numbers*. But this is another story.

Problem 121. What would you suggest as the product of two complex numbers $(2 + 3\sqrt{-1})$ and $(2 - 3\sqrt{-1})$?

Let us finish this section with more difficult problems.

Problem 122. Factor:

(a) $x^4 + 1$;

(b) $x(y^2 - z^2) + y(z^2 - x^2) + z(x^2 - y^2)$;

(c) $a^{10} + a^5 + 1$;

(d) $a^3 + b^3 + c^3 - 3abc$;

(e) $(a + b + c)^3 - a^3 - b^3 - c^3$;

(f) $(a - b)^3 + (b - c)^3 + (c - a)^3$.

Problem 123. Prove that if $a, b > 1$ then $a + b < 1 + ab$.

Hint. Factor $(1 + ab) - (a + b)$.

Problem 124. Prove that if $a^2 + ab + b^2 = 0$ then $a = 0$ and $b = 0$.

Hint. Recall the factorization of $a^3 - b^3$. (Another solution will be discussed later when speaking about quadratic equations.)

Problem 125. Prove that if $a + b + c = 0$ then $a^3 + b^3 + c^3 = 3abc$.

Problem 126. Prove that if

$$\frac{1}{a + b + c} = \frac{1}{a} + \frac{1}{b} + \frac{1}{c}$$

then there are two opposite numbers among a, b, c (i.e. $a = -b$, $a = -c$ or $b = -c$).

33 Rational expressions

One is not allowed to use division in a polynomial (only addition, subtraction, and multiplication). If we allow division too, we get what are called "rational expressions". (The only restriction is that the divisor must not be identically equal to zero.)

Examples:

(a) $\dfrac{ab}{c}$; (b) $\dfrac{a/b}{b/c}$; (c) $\dfrac{1}{1+\dfrac{1}{x}}$; (d) $\dfrac{1}{1+\dfrac{1}{1+\dfrac{1}{1+\dfrac{1}{x}}}}$;

(e) $\dfrac{\dfrac{x}{y}+\dfrac{y}{z}+\dfrac{z}{x}}{\dfrac{y}{x}+\dfrac{z}{y}+\dfrac{x}{z}}+1$; (f) $\dfrac{x^3+x^2+x+1}{x+1}$; (g) $\dfrac{1}{\left(\dfrac{1}{a}+\dfrac{1}{b}\right)/2}$

Let us mention that, for example, $\dfrac{x^2-x^2}{x-x}$ is not a rational expression because the denominator is identically equal to 0.

Let us mention as well that the permisson to use division is not an obligation to use it; therefore, any polynomial is a rational expression.

34 Converting a rational expression into the quotient of two polynomials

A rational expression may include several divisions (as in examples (b), (c), (d), or (g)). But it can be converted into a form in which only one division is used and the division operation is the last one. In other words, any rational expression may be converted into the quotient of two polynomials.

The following transformations are used to do the conversion:

1. Addition: Assume that we want to add $\dfrac{P}{Q}$ and $\dfrac{R}{S}$ where P,Q,R,S are polynomials. Find the common denominator for $\dfrac{P}{Q}$ and $\dfrac{R}{S}$ (if we have no better idea, just multiply P and Q by S and multiply R and S by Q):

$$\frac{P}{Q}+\frac{R}{S}=\frac{PS}{QS}+\frac{QR}{QS}=\frac{PS+QR}{QS}.$$

We've got a quotient of two polynomials.

2. The subtraction case is similar:
$$\frac{P}{Q} - \frac{R}{S} = \frac{PS}{QS} - \frac{QR}{QS} = \frac{PS - QR}{QS}.$$

3. Multiplication:
$$\frac{P}{Q} \cdot \frac{R}{S} = \frac{PR}{QS}.$$

4. Division:
$$\frac{P}{Q} \Big/ \frac{R}{S} = \frac{PS}{QR}.$$

Sometimes during the transformation we are able to simplify the expression, eliminating a common factor in the numerator and the denominator:
$$\frac{PX}{QX} = \frac{P}{Q}.$$

Problem 127. Convert the expressions from the examples (**b**), (**c**), (**d**), (**e**), and (**g**) to this form (expressions (**a**) and (**f**) already are in this form).

Answers and solutions.

(**b**) $\dfrac{ac}{b^2}$; (**c**) $\dfrac{x}{x+1}$;

(**d**) $\dfrac{1}{1 + \dfrac{1}{x}} = \dfrac{x}{x+1}$; $1 + \dfrac{1}{1 + \dfrac{1}{x}} = 1 + \dfrac{x}{x+1} = \dfrac{2x+1}{x+1}$;

$\dfrac{1}{1 + \dfrac{1}{1 + \dfrac{1}{x}}} = \dfrac{x+1}{2x+1}$; $1 + \dfrac{1}{1 + \dfrac{1}{1 + \dfrac{1}{x}}} = \dfrac{3x+2}{2x+1}$;

thus, the answer is $\dfrac{2x+1}{3x+2}$.

(**e**) $\dfrac{\dfrac{x}{y} + \dfrac{y}{z} + \dfrac{z}{x}}{\dfrac{y}{x} + \dfrac{z}{y} + \dfrac{x}{z}} + 1 = \dfrac{(x^2z + y^2x + z^2y)/xyz}{(y^2z + z^2x + x^2y)/xyz} + 1 =$

$= \dfrac{xyz \cdot (x^2z + y^2x + z^2y)/xyz}{xyz \cdot (y^2z + z^2x + x^2y)/xyz} + 1 = \dfrac{(x^2z + y^2x + z^2y)}{(y^2z + z^2x + x^2y)} + 1 =$

$= \dfrac{x^2z + y^2x + z^2y + y^2z + z^2x + x^2y}{y^2z + z^2x + x^2y}$;

(**g**) $\dfrac{2ab}{a+b}$.

Let us mention that in such problems the answer is not defined uniquely. For example, the expression

$$\frac{x^3 + x^2 + x + 1}{x^2 - 1}$$

may be left as is, but may also be transformed as follows:

$$\frac{x^3 + x^2 + x + 1}{(x+1)(x-1)} = \frac{(x+1)(x^2+1)}{(x+1)(x-1)} = \frac{x^2+1}{x-1}$$

Remark. Strictly speaking, the cancellation of common factors is not a perfectly legal operation, because sometimes the factor being cancelled may be equal to zero. For example, $\dfrac{x^3 + x^2 + x + 1}{x^2 - 1}$ is undefined when $x = -1$; it is equal to $\dfrac{x^2+1}{x-1}$ where both are defined. Usually this effect is ignored but sometimes it may become important.

Sometimes the statement of a problem requires us to "simplify the expression" – to convert it to the simplest possible form. Though simplicity is a matter of taste, usually it is clear what the author of the problem meant.

Problem 128. Simplify the expression

$$\frac{(x-a)(x-b)}{(c-a)(c-b)} + \frac{(x-a)(x-c)}{(b-a)(b-c)} + \frac{(x-b)(x-c)}{(a-b)(a-c)}.$$

Solution. Let us first add two fractions. The common denominator is $(c-a)(c-b)(b-a)$. Additional factors are $b-a$ for the first fraction and $c-a$ for the second. We use the fact that $b - c = -(c - b)$, so

$$\frac{(x-a)(x-b)}{(c-a)(c-b)} + \frac{(x-a)(x-c)}{(b-a)(b-c)} =$$

$$= \frac{(x-a)(x-b)(b-a) - (x-a)(x-c)(c-a)}{(c-a)(c-b)(b-a)} =$$

$$= \frac{(x-a)[(x-b)(b-a) - (x-c)(c-a)]}{(c-a)(c-b)(b-a)} =$$

$$= \frac{(x-a)[xb - \underline{xa} - b^2 + ab - xc + \underline{xa} + c^2 - ac]}{(c-a)(c-b)(b-a)} =$$

$$= \frac{(x-a)[x(b-c)+a(b-c)-(b-c)(b+c)]}{(c-a)(c-b)(b-a)} =$$

$$= \frac{(x-a)(b-c)(x+a-b-c)}{(c-a)(c-b)(b-a)}$$

Reducing the common factors $(c-b) = -(b-c)$ we get

$$\frac{(x-a)(b+c-a-x)}{(c-a)(b-a)}$$

Now we can add the third fraction (it has the same denominator, because minus times minus is plus):

$$\frac{(x-a)(b+c-a-x)}{(c-a)(b-a)} + \frac{(x-b)(x-c)}{(a-b)(a-c)} =$$

$$= \frac{xb+xc-xa-x^2-ab-ac+a^2+ax+x^2-xb-xc+bc}{(c-a)(b-a)} =$$

$$= \frac{a^2+bc-ab-ac}{(c-a)(b-a)} = \frac{a(a-b)-c(a-b)}{(c-a)(b-a)} = \frac{(a-c)(a-b)}{(c-a)(b-a)} = 1$$

So we have proved the identity

$$\frac{(x-a)(x-b)}{(c-a)(c-b)} + \frac{(x-a)(x-c)}{(b-a)(b-c)} + \frac{(x-b)(x-c)}{(a-b)(a-c)} = 1.$$

Problem 129. Check this identity in the special cases $x = a$, $x = b$, and $x = c$.

We shall see later that in fact these three cases are sufficient to be sure that the identity is true in the general case. (So the long computations we have done could be avoided.) But we need more theory to realize this.

To conclude this section we state some problems involving rational expressions.

The expression

$$\frac{1}{\left(\dfrac{1}{a}+\dfrac{1}{b}\right)/2}$$

(the inverse of the arithmetic mean of the inverses of a and b; see below) is called the *harmonic mean* of a and b. You may meet it in some situations.

Problem 130. A swimming pool is divided into two equal sections. Each section has its own water supply pipe. To fill one section (using its pipe) you need a hours. To fill the other section you need b hours. How many hours would you need if you turn on both pipes and remove the wall dividing the pool into sections?

Problem 131. A motor boat needs a hours to go from A to B down the river and needs b hours to go from B to A (up the river). How many hours would it need to go from A to B if there were no current in the river?

Problem 132. For the first half of a trip a car has velocity v_1; for the second half of a trip it has the velocity v_2. What is the mean velocity of the car?

Problem 133. You know that $x + \dfrac{1}{x} = 7$. Compute (a) $x^2 + \dfrac{1}{x^2}$; (b) $x^3 + \dfrac{1}{x^3}$.

Problem 134. You know that $x + \dfrac{1}{x}$ is an integer. Prove that $x^n + \dfrac{1}{x^n}$ is an integer for any $n = 1, 2, 3$, etc.

Problem 135. Solving problem (d) on pages 56–57 we have seen that

$$\frac{1}{1 + \dfrac{1}{x}} = \frac{x}{x+1}, \quad \frac{1}{1 + \dfrac{1}{1 + \dfrac{1}{x}}} = \frac{x+1}{2x+1}, \quad \frac{1}{1 + \dfrac{1}{1 + \dfrac{1}{1 + \dfrac{1}{x}}}} = \frac{2x+1}{3x+2}.$$

Represent the fractions

$$\frac{1}{1 + \dfrac{1}{1 + \dfrac{1}{1 + \dfrac{1}{1 + \dfrac{1}{x}}}}}, \quad \frac{1}{1 + \dfrac{1}{1 + \dfrac{1}{1 + \dfrac{1}{1 + \dfrac{1}{x}}}}}, \quad \ldots$$

as quotients of two polynomials and try to find a law governing the coefficients of these polynomials. (These fractions are examples of so-called *continued fractions.* The coefficients of the polynomials in question turn out to be the so-called *Fibonacci numbers*; see page 87)

35 Polynomial and rational fractions in one variable

If a polynomial contains only one variable, its standard form consists of its monomials written in the order of decreasing degrees. The monomial having the highest degree is called the *first* monomial. Its degree is called the degree of the polynomial. (Of course, monomials with zero coefficients must be ignored. For a zero polynomial the degree is undefined.) For example, the polynomial $7x^2 + 3x + 1$ has the first monomial $7x^2$ and degree 2. Constants (not equal to zero) are considered as polynomials of degree 0.

Problem 136. What is the first term of the polynomial $(2x+1)^5$?

Problem 137. Assume that a polynomial P has degree m and the polynomial Q has degree n. Find the degree of their product $P \cdot Q$.

Solution. When multiplying the first monomials we get a monomial of degree $m + n$ (because $x^m \cdot x^n = x^{m+n}$); its coefficient is the product of the coefficients of x^m and x^n in P and Q. This monomial is the only one having degree $m+n$; all the others have smaller degree. So there is nothing to cancel it and thus it will remain after reducing similar terms.

Problem 138. (a) What can be said about the degree of the sum of two polynomials having degrees 7 and 9? (b) What can be said about the degree of the sum of two polynomials both having degree 7?

Answer. (a) It is 9; (b) any degree not exceeding 7 is possible.

Problem 139. Consider a polynomial in one variable x having degree 10. Substitute $y^7 + 5y^2 - y - 4$ for x in this polynomial and get a polynomial in y. What can be said about its degree?

36 Division of polynomials in one variable; the remainder

Common fractions are either *proper* or *improper*. A proper fraction is a fraction where the numerator is smaller than the denominator, such as $\dfrac{3}{7}$ or $\dfrac{1}{15}$. An improper fraction is a fraction where the numerator is not less than the denominator, such as $\dfrac{7}{5}$, $\dfrac{11}{11}$, or $\dfrac{37}{7}$.

Any improper fraction has an integer part, which is obtained when we divide the numerator by the denominator, plus a proper fraction. For example:

$$7 = 1 \cdot 5 + 2$$
(quotient 1, remainder 2)

$$\frac{7}{5} = 1 + \frac{2}{5}$$

$$5\overline{)7}\;^{1\ \text{r. }2}$$

Another example:

$$37 = 5 \cdot 7 + 2$$
(quotient 5, remainder 2)

$$\frac{37}{7} = 5 + \frac{2}{7}$$

$$7\overline{)37}\;^{5\ \text{r. }2}$$

Now an example where the remainder is zero:

$$11 = 1 \cdot 11 + 0$$
(quotient 1, remainder 0)

$$\frac{11}{11} = 1$$

$$11\overline{)11}\;^{1\ \text{r. }0}$$

Now we shall learn to do similar transformations for fractions whose numerators and denominators are polynomials in one variable. Such a fraction is considered proper if the degree of its numerator is less than the degree of its denominator. For example, the fractions

$$\frac{10x}{x^2}, \quad \frac{1}{x^3 - 1}$$

are proper, while the fractions

$$\frac{x^4}{x - 2}, \quad \frac{x + 1}{x + 2}, \quad \frac{x^3}{5x}, \quad \frac{x^3 + 1}{x + 1}$$

are improper.

Any improper fraction can be converted into the sum of a polynomial and a proper fraction. Several examples:

(a) $\dfrac{x + 3}{x + 1} = \dfrac{(x + 1) + 2}{x + 1} = 1 + \dfrac{2}{x + 1}.$

(b) $\dfrac{x}{x+2} = \dfrac{(x+2)-2}{x+2} = 1 - \dfrac{2}{x+2}$.

(c) $\dfrac{x}{2x+1} = \dfrac{x+(1/2)}{2x+1} - \dfrac{1/2}{2x+1} = \dfrac{1}{2} - \dfrac{1/2}{2x+1}$.

(When we say that a polynomial must not contain division it does not mean that all its coefficients must be integers; they may be any numbers, including fractions. So, for example, $\dfrac{1}{2}$ is a perfectly legal polynomial of degree 0.)

(d) $\dfrac{x^2}{x-2} = \dfrac{(x^2-4)+4}{x-2} = \dfrac{(x+2)(x-2)+4}{x-2} = (x+2)+\dfrac{4}{x-2}$.

(e) $\dfrac{x^4}{x-2} = \dfrac{(x^4-16)+16}{x-2} = \dfrac{(x^2+4)(x+2)(x-2)+16}{x-2} =$

$= (x^2+4)(x+2) + \dfrac{16}{x-2}$.

There is a standard way of converting an improper fraction (where the numerator and the denominator are polynomials) into a sum of a polynomial and a proper fraction. It is similar to the division of numbers. Let us illustrate it by examples.

Example. Converting the improper fraction $\dfrac{x^4}{x-2}$:

$$
\begin{array}{r}
x^3 + 2x^2 + 4x + 8 \quad \leftarrow \quad \text{the quotient} \\
\hline
x - 2 \,\big|\, x^4 \\
x^4 - 2x^3 \\
\hline
2x^3 \\
2x^3 - 4x^2 \\
\hline
4x^2 \\
4x^2 - 8x \\
\hline
8x \\
8x - 16 \\
\hline
16 \quad \leftarrow \quad \text{the remainder}
\end{array}
$$

The same procedure can be written in another, less readable, way:

$\dfrac{x^4}{x-2} = \dfrac{x^4-2x^3}{x-2} + \dfrac{2x^3}{x-2} = x^3 + \dfrac{2x^3}{x-2} = x^3 + \dfrac{2x^3-4x^2}{x-2} + \dfrac{4x^2}{x-2} =$

$= x^3 + 2x^2 + \dfrac{4x^2}{x-2} = x^3 + 2x^2 + \dfrac{4x^2-8x}{x-2} + \dfrac{8x}{x-2} =$

$= x^3 + 2x^2 + 4x + \dfrac{8x-16}{x-2} + \dfrac{16}{x-2} = x^3 + 2x^2 + 4x + 8 + \dfrac{16}{x-2}$.

So we get
$$x^4 = (x^3 + 2x^2 + 4x + 8)(x - 2) + 16.$$

Example. Now let us convert the fraction $\dfrac{x^3 + 2x}{x^2 - x + 1}$:

$$
\begin{array}{r}
x + 1 \\
x^2 - x + 1 \overline{\smash{\big)}\, x^3 + 2x } \\
\underline{x^3 - x^2 + x} \\
x^2 + x \\
\underline{x^2 - x + 1} \\
2x - 1
\end{array}
$$

The same conversion written in another way:

$$
\begin{aligned}
\frac{x^3 + 2x}{x^2 - x + 1} &= \frac{x^3 - x^2 + x}{x^2 - x + 1} + \frac{x^2 + x}{x^2 - x + 1} = x + \frac{x^2 + x}{x^2 - x + 1} = \\
&= x + \frac{x^2 - x + 1}{x^2 - x + 1} + \frac{2x - 1}{x^2 - x + 1} = (x + 1) + \frac{2x - 1}{x^2 - x + 1}.
\end{aligned}
$$

So we get
$$x^3 + 2x = (x + 1)(x^2 - x + 1) + (2x - 1).$$

Example. The last example of fraction conversion: $\dfrac{x^3}{2x - 3}$.

$$
\begin{array}{r}
(1/2)x^2 + (3/4)x + (9/8) \\
2x - 3 \overline{\smash{\big)}\, x^3 } \\
\underline{x^3 - (3/2)x^2} \\
(3/2)x^2 \\
\underline{(3/2)x^2 - (9/4)x} \\
(9/4)x \\
\underline{(9/4)x - 27/8} \\
27/8
\end{array}
$$

The same conversion:

$$
\begin{aligned}
\frac{x^3}{2x - 3} &= \frac{x^3 - (3/2)x^2}{2x - 3} + \frac{(3/2)x^2}{2x - 3} = \\
&= \frac{1}{2}x^2 + \frac{(3/2)x^2 - (9/4)x}{2x - 3} + \frac{(9/4)x}{2x - 3}
\end{aligned}
$$

$$= \frac{1}{2}x^2 + \frac{3}{4}x + \frac{(9/4)x - (27/8)}{2x - 3} + \frac{27/8}{2x - 3} =$$
$$= \left(\frac{1}{2}x^2 + \frac{3}{4}x + \frac{9}{8}\right) + \frac{27/8}{2x - 3}.$$

So we get

$$x^3 = \left(\frac{1}{2}x^2 + \frac{3}{4}x + \frac{9}{8}\right)(2x - 3) + \frac{27}{8}$$

Now it is time for the exact definition of polynomial division.

Definition. Assume that we have two polynomials (in one variable), called the dividend and the divisor. To perform a division means to find two other polynomials, called the quotient and the remainder, such that

$$\boxed{(\text{dividend}) = (\text{quotient}) \cdot (\text{divisor}) + (\text{remainder})}$$

where the degree of the remainder is less than the degree of the divisor (or the remainder is zero).

Problem 140. What can you say about the degrees of the remainder and the quotient if a polynomial of degree 7 is divided by a polynomial of degree 3?

Answer. The degree of the quotient is 4; the degree of the remainder may be 0, 1, 2, or 3 or undefined (the remainder may be absent or, rather, equal to zero).

Problem 141. Prove that the quotient and the remainder with the desired properties do exist and are unique.

Solution. In the examples above we have seen a method of finding the quotient and the remainder with the desired properties, so they do exist. Their uniqueness can be proved as follows. Assume that we divide P by S and have two possible quotients Q_1 and Q_2 and two corresponding remainders R_1 and R_2. So we have

$$P = Q_1 S + R_1$$
$$P = Q_2 S + R_2$$

and both R_1 and R_2 have degree less than the degree of S. Then

$$Q_1 S + R_1 = Q_2 S + R_2$$

and, therefore,

$$R_1 - R_2 = Q_2 S - Q_1 S = (Q_1 - Q_2)S.$$

Here $R_1 - R_2$ is a difference of two polynomials of degree smaller than the degree of S, so their difference has degree smaller than the degree of S and cannot be a multiple of S unless it is equal to 0. Therefore, $Q_1 - Q_2 = 0$, that is, $Q_1 = Q_2$, hence, $R_1 = R_2$.

Problem 142. What happens if the degree of the dividend is smaller than the degree of the divisor?

Answer. In this case the fraction is already proper, so the quotient is equal to 0 and the remainder is equal to the dividend.

Polynomial division is similar to ordinary division:

$$
\begin{array}{r}
112 \\
11\overline{)1234} \\
\underline{11} \\
13 \\
\underline{11} \\
24 \\
\underline{22} \\
2
\end{array}
\qquad
\begin{array}{r}
x^2 + x + 2 \\
x+1\overline{)x^3 + 2x^2 + 3x + 4} \\
\underline{x^3 + x^2} \\
x^2 + 3x \\
\underline{x^2 + x} \\
2x + 4 \\
\underline{2x + 2} \\
2
\end{array}
$$

$$1234 = 112 \cdot 11 + 2 \qquad x^3 + 2x^2 + 3x + 4 = (x^2 + x + 2)(x + 1) + 2$$

In this example we have a perfect analogy; to see it, substitute 10 for x in the polynomial division. In other cases such as

$$
\begin{array}{r}
x^2 + 3x + 6 \\
x-1\overline{)x^3 + 2x^2 + 3x + 4} \\
\underline{x^3 - x^2} \\
3x^2 + 3x \\
\underline{3x^2 - 3x} \\
6x + 4 \\
\underline{6x - 6} \\
10
\end{array}
$$

$$x^3 + 2x^2 + 3x + 4 = (x^2 + 3x + 6)(x - 1) + 10$$

the analogy is incomplete; if we substitute 10 for x, we get the equality $1234 = 136 \cdot 9 + 10$, which is true but does not mean that dividing 124 by 9 we get quotient 136 and remainder 10 (in fact, 137 is the quotient and 1 is the remainder).

Problem 143.

(a) Divide $x^3 - 1$ by $x - 1$;

(b) Divide $x^4 - 1$ by $x - 1$;

(c) Divide $x^{10} - 1$ by $x - 1$;

(d) Divide $x^3 + 1$ by $x + 1$;

(e) Divide $x^4 + 1$ by $x + 1$.

Problems (a)–(c) are special cases of the formula

$$\frac{x^n - 1}{x - 1} = x^{n-1} + x^{n-2} + \cdots + x^2 + x + 1$$

which can be easily checked by division (as described) or just by multiplication of $x - 1$ and $x^{n-1} + x^{n-2} + \cdots + x^2 + x + 1$. This formula can also be considered as a way to compute the sum of consecutive powers of a number x:

$$1 + x + x^2 + \cdots + x^{n-1} = \frac{x^n - 1}{x - 1}$$

(it is valid for all x except 1). See below about the sum of a geometric progression.

Problem 144. The powers of 2,

$$1, 2, 4, 8, 16, 32, 64, \ldots$$

have the following property: The sum of all members of this sequence up to any term is 1 less than the next term; for example

$$
\begin{aligned}
1 + 2 &= & 3 = 4 - 1 \\
1 + 2 + 4 &= & 7 = 8 - 1 \\
1 + 2 + 4 + 8 &= & 15 = 16 - 1
\end{aligned}
$$

and so on. Explain why.

Solution. Look at the equation

$$1 + x + x^2 + \cdots + x^{n-1} = \frac{x^n - 1}{x - 1}$$

when $x = 2$. We get

$$1 + 2 + 2^2 + \cdots + 2^{n-1} = \frac{2^n - 1}{2 - 1} = 2^n - 1.$$

Another solution. To compute the sum $1 + 2 + 4 + 8 + 16$, let us add and subtract 1:

$$
\begin{aligned}
1 + 2 + 4 + 8 + 16 &= \\
&= (1 + 1 + 2 + 4 + 8 + 16) - 1 = \\
&= (2 + 2 + 4 + 8 + 16) - 1 = \\
&= (4 + 4 + 8 + 16) - 1 = \\
&= (8 + 8 + 16) - 1 = \\
&= (16 + 16) - 1 = \\
&= 32 - 1.
\end{aligned}
$$

37 The remainder when dividing by $x - a$

There is a method that allows us to find the remainder of an arbitrary polynomial divided by $x - a$ without actually performing the division.

Assume that we want to find the remainder when x^4 is divided by $x - 2$. We can be sure that the remainder is a number (its degree must be less than the degree of $x - 2$). To find this number, look at the equality

$$x^4 = (\text{quotient})(x - 2) + (\text{remainder})$$

and substitute $x = 2$. We get

$$2^4 = (\ldots) \cdot 0 + (\text{remainder});$$

so the remainder is equal to $2^4 = 16$.

In general, if P is an arbitrary polynomial that we want to divide by $x - a$ (where a is some number), we write

$$P(x) = (\text{quotient})(x - a) + (\text{remainder})$$

and substitute a for x. Therefore,

To find the remainder when P is divided by $x-a$, substitute a for x in P.

This rule is called the *remainder theorem*, or Bezout's theorem. It allows us to find the remainder without the actual division. However, if you want to know the quotient, you need to perform the division.

Here is a consequence of Bezout's theorem:

To find whether a polynomial P is divisible by $x - a$ (without remainder), test whether it becomes zero after substitution of a for x.

If a polynomial P becomes zero when some number a is substituted for x, then this number a is called a *root* of the polynomial P. Therefore we may say

P is divisible by $x - a$ $\iff$ a is a root of P.

Problem 145. (a) For which n is the polynomial $x^n - 1$ divisible by $x-1$? (b) For which n is the polynomial $x^n + 1$ divisible by $x+1$?

After we find a root of a polynomial we may factor it; $x - a$ is one of the factors. Then we may try to apply the same method to the quotient.

Problem 146. Factor these polynomials:

(a) $x^4 + 5x - 6$;

(b) $x^4 + 3x^2 + 5x + 1$;

(c) $x^3 - 3x - 2$.

Problem 147. The numbers 1 and 2 are roots of a polynomial P. Prove that P is divisible by $(x - 1)(x - 2)$.

Solution. P is divisible by $x-1$ because 1 is a root of P. Therefore $P = (x - 1) \cdot Q$ for some polynomial Q. Substituting 2 for x in this equality we find that 2 is a root of Q, so Q is divisible by $x - 2$, that is, $Q = (x - 2) \cdot R$ for some polynomial R. So $P = (x - 1)(x - 2)R$.

Remark. A typical wrong solution goes as follows: P is divisible by $x - 1$ (because 1 is a root) and by $x - 2$ (because 2 is a root), therefore P is divisible by $(x-1)(x-2)$. The error: "therefore" is not justified. For example, 12 is divisible by 6 and by 4, but we may not say "therefore, 12 is divisible by $6 \cdot 4 = 24$".

A similar argument shows that

> If different numbers $a_1, a_2, \ldots, a_n$ are roots of a polynomial P, then P is divisible by $(x - a_1)(x - a_2) \cdots (x - a_n)$.

Problem 148. What is the maximal possible number of roots for a polynomial having degree 5?

Solution. The answer is 5. For example, the polynomial

$$(x - 1)(x - 2)(x - 3)(x - 4)(x - 5)$$

has 5 roots. More than 5 roots is impossible; if a polynomial P had 6 roots $a_1, a_2, a_3, a_4, a_5, a_6$, then it would be divisible by

$$(x - a_1)(x - a_2) \cdots (x - a_6),$$

that is,

$$P = (x - a_1)(x - a_2) \cdots (x - a_6)Q$$

for some Q. That is impossible because the degree of the right-hand side is at least 6.

In general, a polynomial of degree n may have at most n different roots.

Remark. We used here the expression "different roots" because the words "the number of roots" may be used in a different sense. For example, what is the number of roots of the polynomial $x^2 - 2x + 1$? The polynomial is equal to $(x - 1)^2$, so $x = 1$ is a root and all $x \neq 1$ are not roots. So we may say that it has exactly one root. On the other hand, the general formula for a polynomial with two roots a and b is

$$c(x - a)(x - b)$$

and our polynomial

$$x^2 - 2x + 1 = (x - 1)^2 = (x - 1)(x - 1)$$

is a special case of this formula when $a = b = 1$ (and $c = 1$); so mathematicians often say that this polynomial has "two equal roots".

We shall not use this terminology in this book but you may see it, for example, in the statement of the so-called "fundamental theorem of algebra" claiming that "any polynomial of degree n has exactly n complex roots"

Problem 149. How should you check whether a given polynomial P is divisible by $x^2 - 1$?

Answer. Check whether 1 and -1 are roots of P.

Problem 150. For which n is the polynomial $x^n - 1$ divisible by $x^2 - 1$?

Now let us return to the identity

$$\frac{(x-a)(x-b)}{(c-a)(c-b)} + \frac{(x-a)(x-c)}{(b-a)(b-c)} + \frac{(x-b)(x-c)}{(a-b)(a-c)} - 1 = 0$$

which we discussed on page 59 (we have moved 1 to the left-hand side of the equation). Assume that a, b, c are different numbers. Consider the left-hand side as a polynomial with one variable x. The degree of this polynomial does not exceed 2. Therefore it may have at most two roots (if it is not equal to zero). But a, b, and c are its roots! Therefore, it is equal to zero.

A careful reader would say that we made a big mistake: we confuse the equality of rational expressions for all numerical values of a, b, c, x (strictly speaking, not even for all, because, for example, the left-hand side is undefined when $a = b$) with the possibility of transforming the left-hand side to zero according to algebraic rules. What can be said about this? Bad news: this really is a problem. Good news: this problem is not fatal (but to justify this transition you need some theory).

Problem 151. The remainder of a polynomial P (in one variable x) when divided by $x^2 - 1$ is a polynomial of degree at most 1, that is, it has the form $ax + b$ for some numbers a and b. How can you find a and b if you know the values of P when $x = -1$ and $x = 1$?

Hint. Look at the equality

$$P = (x^2 - 1)(\text{quotient}) + (ax + b)$$

71

and substitute 1 and -1 for x.

Problem 152. The polynomial P gives a remainder of $5x - 7$ when divided by $x^2 - 1$. Find the remainder when P is divided by $x - 1$.

Problem 153. The polynomial $P = x^3 + x^2 - 10x + 1$ has three different roots (the authors guarantee it) denoted by x_1, x_2, x_3. Write a polynomial with integer coefficients having roots

(a) $x_1 + 1, x_2 + 1, x_3 + 1$; (b) $2x_1, 2x_2, 2x_3$; (c) $\dfrac{1}{x_1}, \dfrac{1}{x_2}, \dfrac{1}{x_3}$.

Problem 154. Assume that $x^3 + ax^2 + x + b$ (where a and b are some numbers) is divisible by $x^2 - 3x + 2$. Find a and b.

38 Values of polynomials, and interpolation

Assume that a polynomial P includes only one letter (variable) x. To stress this we denote this polynomial by $P(x)$ ("P of x"). Substitute some number, say 6, for x in P and perform all computations. We get a number. This number is called the value of the polynomial P for $x = 6$ and is denoted by $P(6)$ ("P of 6").

For example, if $P(x) = x^2 - x - 4$ then $P(0) = 0^2 - 0 - 4 = -4$. Other values are $P(1) = -4$, $P(2) = -2$, $P(3) = 2$, $P(4) = 8$, $P(5) = 16$, $P(6) = 26$, etc.

Problem 155. Calculate a table of values $P(0), \ldots, P(6)$ for the polynomial $P(x) = x^3 - 2$.

Problem 156. Let us write the values $P(0), P(1), P(2), \ldots$ for $P(x) = x^2 - x - 4$:

$$-4, \ -4, \ -2, \ 2, \ 8, \ 16, \ 26, \ \ldots$$

Under any two adjacent numbers write their difference:

-4		-4		-2		2		8		16		26 $\ldots$
	0		2		4		6		8		10 $\ldots$	

and repeat the same operation with this sequence of "first differences":

-4		-4		-2		2		8		16		26 $\ldots$
	0		2		4		6		8		10 $\ldots$	
		2		2		2		2		2 $\ldots$		

Now all numbers are 2 s. Prove that it is not a coincidence and that all subsequent numbers (called "second differences") are also 2 s.

Problem 157. Prove that for any polynomial of degree 2 all second differences are equal.

Problem 158. What can be said about polynomials having degree 3?

Problem 159. (L. Euler) Compute the values $P(x) = x^2 + x + 41$ for $x = 1, 2, 3, \ldots, 10$. Check that all these values are prime numbers (having no divisors except 1 and themselves). Might it be that all of $P(1), P(2), P(3), \ldots$ are prime numbers for this polynomial P?

Now we address the following topic: What can be said about a polynomial if we have some information about its values?

By a polynomial of degree *not exceeding* n we mean any polynomial of degree n, $n - 1, \ldots, 2$, 1, 0, or the zero polynomial (whose degree is undefined).

For example, the general form of a polynomial of degree not exceeding 1 is $ax + b$. When $a \neq 0$ it has degree 1. When $a = 0, b \neq 0$ it has degree 0. When $a = b = 0$ we get the zero polynomial whose degree is undefined.

In a similar way the general form of a polynomial of degree not exceeding 2 is $ax^2 + bx + c$, etc.

Problem 160. You know that $P(x)$ is a polynomial of degree not exceeding 1, that $P(1) = 7$, and that $P(2) = 5$. Find $P(x)$.

Solution. By definition, $P(x) = ax + b$, where a and b are some numbers. Let us substitute $x = 1$ and $x = 2$. We get:

$$
\begin{aligned}
P(1) &= a + b &= 7 \\
P(2) &= 2a + b &= 5
\end{aligned}
$$

Comparing this equations we see that after adding one more a (in the second one), 7 becomes 5, so $a = -2$. Therefore $b = 9$. Answer: $P(x) = -2x + 9$.

The same method can be applied to find a polynomial of degree not exceeding 1 if we know its values for any two different values of x.

If you know that the graph of a function $y = ax + b$ is a straight line you can easily explain this fact geometrically; for any two points there is exactly one straight line going through these points. (Two given values for two values of x correspond to two points in the plane.)

Problem 161. A polynomial $P(x)$ of degree not exceeding 1 satisfies the conditions $P(1) = 0$, $P(2) = 0$. Prove that $P(x) = 0$ for any x.

Now we consider polynomials of degree not exceeding 2. How many values do we need to reconstruct such a polynomial? We shall see that two is not enough.

Problem 162. A polynomial $P(x)$ of degree not exceeding 2 satisfies the conditions $P(1) = 0$, $P(2) = 0$. Can we conclude that $P(x) = 0$?

Solution. No; look at the polynomial $P(x) = (x - 1)(x - 2) = x^2 - 3x + 2$.

We already know that any polynomial $P(x)$ such that $P(1) = P(2) = 0$ has the form $P(x) = (x - 1)(x - 2)Q(x)$ where $Q(x)$ is some polynomial. If we also know that $P(x)$ has degree not exceeding 2, then $Q(x)$ must be a number (otherwise the degree of P will be too big).

Problem 163. A polynomial $P(x)$ of degree not exceeding 2 satisfies the conditions $P(1) = 0$, $P(2) = 0$, $P(3) = 4$. Find $P(x)$.

Solution. As we have seen, $P(x) = a(x - 1)(x - 2)$ where a is some constant. To find a, substitute $x = 3$:

$$P(3) = a(3 - 1)(3 - 2) = 2a = 4;$$

therefore $a = 2$. Answer: $P(x) = 2(x - 1)(x - 2) = 2x^2 - 6x + 4$.

Another solution. Any polynomial of degree not exceeding 2 has the form $ax^2 + bx + c$. Substituting $x = 1$, $x = 2$ and $x = 3$, we get

$$\begin{array}{rcccl} P(1) & = & a + b + c & = & 0 \\ P(2) & = & 4a + 2b + c & = & 0 \\ P(3) & = & 9a + 3b + c & = & 4. \end{array}$$

Therefore,

$$\begin{array}{rcccl} P(2) - P(1) & = & 3a + b & = & 0 \\ P(3) - P(2) & = & 5a + b & = & 4. \end{array}$$

74

Additional $2a$ make 4 from 0, therefore $a = 2$. Now we can find $b = -6$ and then $c = 4$. Answer: $2x^2 - 6x + 4$.

Problem 164. Prove that a polynomial of degree not exceeding 2 is defined uniquely by three of its values.

This means that if $P(x)$ and $Q(x)$ are polynomials of degree not exceeding 2 and $P(x_1) = Q(x_1)$, $P(x_2) = Q(x_2)$, $P(x_3) = Q(x_3)$ for three different numbers x_1, x_2, and x_3, then the polynomials $P(x)$ and $Q(x)$ are equal.

Solution. Consider a polynomial $R(x) = P(x) - Q(x)$. Its degree does not exceed 2. On the other hand, we know that

$$R(x_1) = R(x_2) = R(x_3) = 0;$$

in other words, x_1, x_2, and x_3 are roots of the polynomial $R(x)$. But a polynomial of degree not exceeding 2, as we know, cannot have more than 2 roots, unless it is equal to zero. Therefore $R(x)$ is equal to zero and $P(x) = Q(x)$.

Problem 165. Assume that

$$16a + 4b + c = 0$$
$$49a + 7b + c = 0$$
$$100a + 10b + c = 0.$$

Prove that $a = b = c = 0$.

Problem 166. Prove that a polynomial of degree not exceeding n is defined uniquely by its $n + 1$ values. (We have already solved this problem for $n = 2$.)

Problem 167. Find a polynomial $P(x)$ of degree not exceeding 2 such that

(a) $P(1) = 0$, $P(2) = 0$, $P(3) = 4$;

(b) $P(1) = 0$, $P(2) = 2$, $P(3) = 0$;

(c) $P(1) = 6$, $P(2) = 0$, $P(3) = 0$;

(d) $P(1) = 6$, $P(2) = 2$, $P(3) = 4$.

Solution. Problem **(a)** was already solved, and the answer was $2(x-1)(x-2)$. Problems **(b)** and **(c)** may be solved by the same method; the answers are $-2(x-1)(x-3)$ for **(b)** and $3(x-2)(x-3)$ for **(c)**. Now we are able to solve **(d)** by just adding the three polynomials from **(a)**, **(b)**, and **(c)**. We get the following answer:

$$2(x-1)(x-2) - 2(x-1)(x-3) + 3(x-2)(x-3) =$$
$$= \quad 2x^2 - 6x + 4 - 2x^2 + 8x - 6 + 3x^2 - 15x + 18 =$$
$$= \quad 3x^2 - 13x + 16.$$

Another solution for (d). Let us find any polynomial Q of degree not exceeding 2 such that $Q(1) = 6$ and $Q(2) = 2$. For example, the polynomial $Q(x) = 10 - 4x$ (having degree 1) will work. It has two desired values $Q(1)$ and $Q(2)$, but unfortunately the value $Q(3)$ is not what we want: $Q(3) = -2$ (and we want 4). The remedy: consider a polynomial

$$P(x) = Q(x) + a(x-1)(x-2).$$

Any a would not damage the values $P(1) = Q(1) = 6$, $P(2) = Q(2) = 2$. And a suitable a will make $P(3)$ correct:

$$P(3) = Q(3) + 2a.$$

To get $P(3) = 4$ we use $a = 3$. So the answer is

$$P(x) = 10 - 4x + 3(x-1)(x-2) =$$
$$= \quad 10 - 4x + 3x^2 - 9x + 6 = 3x^2 - 13x + 16.$$

Problem 168. Find a polynomial $P(x)$ of degree not exceeding 3 such that $P(-1) = 2$, $P(0) = 1$, $P(1) = 2$, $P(2) = 7$.

Problem 169. Assume that $x_1, \ldots, x_{10}$ are different numbers, and $y_1, \ldots, y_{10}$ are arbitrary numbers. Prove that there is one and only one polynomial $P(x)$ of degree not exceeding 9 such that $P(x_1) = y_1$, $P(x_2) = y_2, \ldots, P(x_{10}) = y_{10}$.

Problem 170. Without any computations prove that there exist numbers a, b, and c such that

$$100a + 10b + c \quad = \quad 18.37$$
$$36a + 6b + c \quad = \quad 0.05$$
$$4a + 2b + c \quad = \quad -3$$

(You don't need to find these a, b, and c; it is enough to prove that they exist.)

Problem 171. The highest coefficient of $P(x)$ is 1, and we know that $P(1) = 0$, $P(2) = 0$, $P(3) = 0,\ldots$, $P(9) = 0$, $P(10) = 0$. What is the minimal possible degree of $P(x)$? Find $P(11)$ for this case.

Answer. The minimal degree is 10 and $P(11)$ is 3628800 in this case.

39 Arithmetic progressions

In the sequence of numbers

$$3,\ 5,\ 7,\ 9,\ 11,\ \ldots$$

each term is greater than the preceding one by two units. In the sequence

$$10,\ 9,\ 8,\ 7,\ 6,\ \ldots$$

each term is one unit smaller than the preceding one. Such sequences are called *arithmetic progressions*. (Here "e" is stressed: arithmEtic, not arIthmetic!)

Definition. An *arithmetic progression* is a sequence of numbers where each term is a sum of the preceding one and a fixed number. This fixed number is called the *common difference*, or simply *difference*, of the arithmetic progression.

Problem 172. What are the differences in the examples above?

Answer. 2 and -1.

Problem 173. Find the third term of an arithmetic progression

$$5,\ -2,\ \ldots$$

Answer. -9.

Problem 174. Find the 1000th term of an arithmetic progression

$$2,\ 3,\ 4,\ 5,\ 6,\ \ldots$$

Solution. If the progression were

$$1,\ 2,\ 3,\ 4,\ 5,\ \ldots$$

the first term would be 1, the second term would be 2, ..., the 1000th term would be 1000. In our progression, each term is one unit bigger. So the answer is 1001.

Problem 175. Find the 1000th term of the progression

$$2, \ 4, \ 6, \ 8, \ \ldots$$

Problem 176. Find the 1000th term of the progression

$$1, \ 3, \ 5, \ 7, \ \ldots$$

Problem 177. The first term of a progression is a, its difference is d. Find the 1000th term of the progression. Find its nth term.

Solution.

1st term	a
2nd term	$a + d$
3rd term	$a + 2d$
4th term	$a + 3d$
5th term	$a + 4d$
...	...
1000th term	$a + 999d$
...	...
nth term	$a + (n-1)d$

Problem 178. An arithmetic progression with difference d is rewritten in the reverse order, from right to left. Do we get an arithmetic progression? If so, what is its difference?

Problem 179. In an arithmetic progression whose difference is d every second term is deleted. Do we get an arithmetic progression? If so, what is its difference?

Problem 180. The same question if every *third* term is deleted.

Problem 181. The first term of an arithmetic progression is 5, the third term is 8. Find the second term.

Answer. 6.5.

Problem 182. The first term of an arithmetic progression is a, the third term is b. Find the second term.

Answer. $(a + b)/2$.

Problem 183. The first term of an arithmetic progression is a, the 4th term is b. Find the second and the third terms.

Problem 184. Consider the progression

$$1, \ 3, \ 5, \ 7, \ \ldots, \ 993, \ 995, \ 997, \ 999 \, .$$

How many terms does it have?

Hint. The nth term is equal to $2n-1$. (Another way is to compare it with the progression $2, \ 4, \ 6, \ \ldots, 1000$.)

40 The sum of an arithmetic progression

Problem 185. Compute the sum

$$1 + 3 + 5 + 7 + \cdots + 999 \, .$$

Solution. First of all we have to find out how many terms are in this sum (see the problem above). The nth term of this progresson is equal to $1 + (n-1) \cdot 2 = 2n - 2 + 1 = 2n - 1$. It is equal to 999 when $n = 500$. So this progression contains 500 terms. Let us combine them into 250 pairs:

$$(1 + 999) + (3 + 997) + \cdots + (499 + 501) \, .$$

The sum of each pair is equal to 1000. Thus, the answer is 250,000.

Problem 186. The first term of a progression containing n terms is a, its last (nth) term is b. Find the sum of its terms.

Solution. Grouping terms into pairs (as in the preceding problem) we get $n/2$ pairs, and the sum of each pair is $a + b$. Thus, the answer is $\dfrac{n(a + b)}{2}$.

Problem 187. There is an error in the solution of the preceding problem (however, the answer is valid). Find and correct this error.

Solution. All is O.K. if n is even. But when n is odd, the middle term remains unpaired. To avoid the distinction between odd and even numbers of terms, we apply the following trick. Assume that the sum in question is

$$S = 3 + 5 + 7 + 9 + 11 \, .$$

Rewrite it in the reverse order:

$$S = 11 + 9 + 7 + 5 + 3.$$

Now we add these two equalities:

$$2S = \quad 3 \quad +5 \quad +7 \quad +9 \quad +11 \quad +$$
$$+\,11 \quad +9 \quad +7 \quad +5 \quad +\,3$$

and find out that in each column we have two numbers whose sum is 14:

$$3 + 11 = 5 + 9 = 7 + 7 = 9 + 5 = 11 + 3 = 14.$$

So $2S = 5 \cdot 14 = 70$, $S = \dfrac{5 \cdot 14}{2} = 35.$

In the general case we have n columns with the same sum equal to the sum of the first and the last terms, that is, $a + b$. Therefore,

$$S = \frac{n(a + b)}{2}.$$

This argument can be illustrated by a picture. The sum $3 + 5 + 7 + 9 + 11$ can be drawn as

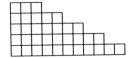

Two such pieces form a rectangle 5×14:

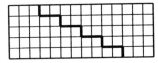

Problem 188. Prove that the sum of n first odd numbers is a perfect square ($1 = 1^2$, $1 + 3 = 2^2$, $1 + 3 + 5 = 3^2$ etc.)

Hint. You may use the preceding problem or the following picture:

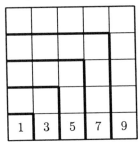

41 Geometric progressions

In the sequence of numbers

$$3,\ 6,\ 12,\ 24,\ \ldots$$

each term is two times bigger than the preceding one. In the sequence

$$6,\ 2,\ \frac{2}{3},\ \frac{2}{9},\ \frac{2}{27}, \ldots$$

each term is three times smaller than the preceding one. Such sequences are called *geometric progressions*.

Definition. A *geometric progression* is a sequence of numbers where each term is a product of the preceding one and a fixed number. This fixed number is called the *common ratio* (or *ratio*) of the geometric progression.

Problem 189. Find the common ratios of the progressions shown above.

Answer. 2, 1/3.

Problem 190. Find the third term of the geometric progression

$$2, 3, \ldots$$

Answer. 9/2.

Problem 191. Find the 1000th term of the geometric progression $3, 6, 12, \ldots$

Solution.

$$
\begin{array}{rccl}
\text{1st term} & 3 & = & 3 \cdot 2^0 \\
\text{2nd term} & 6 & = & 3 \cdot 2^1 \\
\text{3rd term} & 12 & = & 3 \cdot 2^2 \\
& \cdots & & \cdots \\
\text{1000th term} & & = & 3 \cdot 2^{999}
\end{array}
$$

Problem 192. Find the 1000th term of a geometric progression whose first term is a and whose common ratio is q.

Solution.

$$1\text{st term} = a = a \cdot q^0$$
$$2\text{nd term} = a \cdot q = a \cdot q^1$$
$$3\text{rd term} = a \cdot q^2$$
$$4\text{th term} = a \cdot q^3$$
$$\ldots \qquad \ldots$$
$$n\text{th term} = a \cdot q^{n-1}$$

Problem 193. The first term of a geometric progression is 1, the third term is 4. Find the second term. Is your answer the only possible one?

Answer. There are two possibilities: 2 and -2.

Problem 194. A bacterium dividing one a minute fills a vessel in 30 minutes. How much time is necessary for two bacteria to fill the same vessel?

Let us look at the sequence

$$1, 0, 0, 0, \ldots$$

Is it a geometric progression or not? According to our definition it is – each term is equal to the preceding one multiplied by zero (and there is no requirement for the common ratio to be nonzero). Though this sequence looks strange, we do consider it as a geometric progression. (But in some cases we would require the common ratio of a progression to be nonzero.)

Problem 195. A geometric progression whose common ratio is $q \neq 0$ is rewritten in the reverse order, from right to left. Do we get a geometric progression? If so, what is its common ratio?

Answer. $1/q$.

Problem 196. In a geometric progression whose common ratio is q every second term is deleted. Do we get a geometric progression? If so, what is its common ratio?

Problem 197. The same question if every *third* term is deleted.

Problem 198. The first term of a geometric progression is a and the third term is b. Find the second term.

Solution. Assume that x is the second term. Then the common ratio is equal to x/a and at the same time to b/x. Therefore, $x/a = b/x$; multiplying this equality by ax, we get $x^2 = ab$. Therefore, if $ab < 0$ the problem has no solutions (such a progression does not exist); if $ab = 0$ then $x = 0$; if $ab > 0$ there are two possibilities: $x = \sqrt{ab}$ and $x = -\sqrt{ab}$ (see below about square roots).

Remark. Our solution is not applicable when $x = 0$ or $a = 0$. But our formula turns out to be more clever than we may expect. For example, if $a = 1$, $b = 0$ then our formula gives the correct answer $x = \sqrt{1 \cdot 0} = 0$.

Problem 199. The first term of a geometric progression is 1, and its fourth term is $a > 0$. Find the second and the third terms of this progression.

Hint. See below about cube roots.

Answer. $\sqrt[3]{a}$, $\sqrt[3]{a^2}$.

42 The sum of a geometric progression

Problem 200. Compute the sum $1 + 2 + 4 + 8 + \cdots + 512 + 1024$ (each term is twice the preceding one).

Solution. Let us add 1 to this sum:

$$
\begin{aligned}
1 + 1 + 2 + 4 + 8 + \cdots + 1024 \ &= \\
= 2 + 2 + 4 + 8 + \cdots + 1024 \ &= \\
= 4 + 4 + 8 + \cdots + 1024 \ &= \\
= 8 + 8 + \cdots + 1024 \ &= \\
= 16 + \cdots + 1024 \ &= \\
\cdots \ & \\
= 256 + 256 + 512 + 1024 \ &= \\
= 512 + 512 + 1024 \ &= \\
= 1024 + 1024 \ &= \\
= 2048 &
\end{aligned}
$$

So the answer is $2048 - 1 = 2047$.

Another solution. Let us denote this sum by S. Then

$$S = 1 + 2 + 4 + 8 + \cdots + 512 + 1024$$

and

$$2S = 2 + 4 + 8 + 16 + \cdots + 1024 + 2048.$$

The latter sum (compared with the first one) contains an extra term 2048 but does not contain the term 1. So the difference is

$$2S - S = 2048 - 1.$$

Therefore, $S = 2048 - 1 = 2047$.

Problem 201. Compute the sums $1 + \dfrac{1}{2}$, $1 + \dfrac{1}{2} + \dfrac{1}{4}$, $1 + \dfrac{1}{2} + \dfrac{1}{4} + \dfrac{1}{8}$, $\ldots$, $1 + \dfrac{1}{2} + \dfrac{1}{4} + \cdots + \dfrac{1}{1024}$.

Problem 202. The first term of a geometric progression is a and its common ratio is q. Find the sum of the first n terms of this progression.

Solution. The sum in question is equal to

$$a + aq + aq^2 + \cdots + aq^{n-1} = a(1 + q + q^2 + \cdots + q^{n-1}).$$

Recalling the factorization

$$q^n - 1 = (q - 1)(q^{n-1} + q^{n-2} + \cdots + q + 1)$$

we find out that

$$1 + q + \cdots + q^{n-1} = \frac{q^n - 1}{q - 1},$$

so the sum in question is equal to

$$a\frac{q^n - 1}{q - 1}.$$

Another solution. Let us denote the sum in question by S:

$$S = a + aq + \cdots + aq^{n-2} + aq^{n-1}.$$

Multiply it by q:

$$qS = aq + aq^2 + \cdots + aq^{n-1} + aq^n.$$

A new term aq^n appeared and term a disappeared, so

$$qS - S = aq^n - a$$
$$(q - 1)S = a(q^n - 1)$$
$$S = a\frac{q^n - 1}{q - 1}.$$

Problem 203. The solution of the preceding problem has a gap; find it.

Solution. When $q = 1$ the answer given above is absurd; the quotient

$$\frac{1^n - 1}{1 - 1}$$

is undefined. In this case all terms of the progression are equal and the sum is equal to na. So one could say that in a sense

$$\frac{q^n - 1}{q - 1} = n, \quad \text{if } q = 1.$$

(This is a joke, of course – but it is also the computation of the derivative of the function $f(x) = x^n$ from the calculus textbooks!)

43 Different problems about progressions

Problem 204. Is it possible that numbers $1/2$, $1/3$, and $1/5$ are (not necessarily adjacent) terms of the same arithmetic progression?

Hint. Yes. Try $1/30$ as a difference.

Problem 205. Is it possible that the numbers 2, 3, and 5 are (not necessarily adjacent) terms of a geometric progression?

Solution. No, it is impossible. Assume that the common ratio of this progression is equal to q. Then

$$3 = 2q^n, \quad 5 = 3q^m$$

for some m and n. So we get

$$q^n = \frac{3}{2}, \quad q^m = \frac{5}{3}$$

and

$$\left(\frac{3}{2}\right)^m = q^{mn} = \left(\frac{5}{3}\right)^n.$$

Therefore,
$$\frac{3^m}{2^m} = \frac{5^n}{3^n}$$
and $3^{m+n} = 2^m \cdot 5^n$. The left-hand side is an odd number, and the right-hand side is an even number if $m \neq 0$. Hence, m must be equal to 0. But this is also impossible because in this case we would get
$$5 = 3q^m = 3 \cdot 1 = 3 .$$

So we get a contradiction showing that the requirements $3 = 2q^n$ and $5 = 3q^m$ are inconsistent. Hence 2, 3, and 5 could not be terms of the same progression.

Problem 206. In this argument we assumed that the numbers 2, 3, and 5 occur in the progression in this order (because we assume implicitly that m and n are positive integers). What should we do in other cases?

Problem 207. Is it possible that two first terms of an arithmetic progression are integers, but all succeeding terms are not?

Solution. This is impossible; if two adjacent terms are integers, then the difference of the progression is an integer, and all the other terms are also integers.

Problem 208. Is it possible that the first 10 terms of a geometric progression are integers, but all succeeding terms are not?

Solution. Yes, it is possible:
$$512,\ 256,\ 128,\ 64,\ 32,\ 16,\ 8,\ 4,\ 2,\ 1,\ \frac{1}{2},\ \frac{1}{4},\ \ldots$$

Problem 209. Is it possible that the second term of an arithmetic progression is less than its first term and also less than its third term?

Solution. No, in this case the difference of the progression would be positive and negative at the same time.

Problem 210. The same question for a geometric progression.

Solution. Yes, for example, in the progression $1, -1, 1$.

Problem 211. Is it possible that an infinite arithmetic progression contains exactly one integer term?

Hint. Consider the progression with first term 0 and difference $\sqrt{2}$ and use the fact that $\sqrt{2}$ is an irrational number (see below).

Problem 212. Is it possible that an infinite arithmetic progression contains exactly two integer terms?

Answer. No.

Problem 213. In the sequence

$$1, \ 3, \ 7, \ 15, \ 31, \ \dots$$

each term is equal to $2 \times$ (the preceding term) $+ 1$. Find the 100th term of this sequence.

Answer. $2^{100} - 1$.

Problem 214. In a geometric progression each term is equal to the sum of two preceding terms. What can be said about the common ratio of this progression?

Hint. See below about quadratic equations.

Answer. There are two possible common ratios:

$$\frac{1 + \sqrt{5}}{2} \quad \text{and} \quad \frac{1 - \sqrt{5}}{2}.$$

Problem 215. The *Fibonacci sequence*

$$1, \ 1, \ 2, \ 3, \ 5, \ 8, \ 13, \ 21, \ \dots$$

is defined as follows: The first two terms are equal to 1, and each subsequent term is equal to the sum of the two preceding terms. Find numbers A and B such that (for all n), the nth term of the Fibonacci sequence is equal to

$$A\left(\frac{1 + \sqrt{5}}{2}\right)^n + B\left(\frac{1 - \sqrt{5}}{2}\right)^n$$

44 The well-tempered clavier

A musical sound (a tone) consists of air oscillations (produced by string oscillations if we have a string instrument such as a violin or a piano).

The number of oscillations per second is called the *frequency* of oscillations. For example, the note A (*la*) of the fourth octave of a piano has a frequency of 440 oscillations per second (according to the modern standard; the frequency was lower in the past). The higher the tone is, the greater is its frequency.

When we hear two tones together, they form an *interval* (as musicians say). This interval may be consonant (harmonious, nice to hear) or dissonant (not so nice) – or something in between. It turns out that this depends on the ratio of frequencies of the two tones forming the interval. The rule is as follows: A consonant interval appears when the frequency ratio is equal (or very close) to the ratio of two small integers.

For example, an *octave* interval appears when the ratio is equal to $2 = 2/1$. The notes forming an octave interval have the same name. For example, all notes with frequencies 440, 880, 1760, and so on – as well as notes with frequencies 220, 110, and so on – share the name A (*la*), but belong to "different octaves", as musicians say. The octave interval opens the "Campanella" (the final part of the second violin concerto by Paganini; both tones are F-sharp).

A *fifth* is an interval whose frequencies ratio is equal to $3/2$ (or very close to $3/2$; see below). The adjacent strings on a violin form such an interval (G–D, D–A, or A–E). The final part of the Brahms concerto for violin and cello (A minor) starts with a fifth (A–E) played by the cello.

The interval with frequency ratio $4/3$ is called a *fourth*, the interval with ratio $5/4$ is called a *major third*, and the interval with ratio $6/5$ is called a *minor third*.

Problem 216. In a three-tone melody the first two tones form a fifth (the second tone is lower), and the next two tones form a fourth (the last tone is lower). What is the interval between the first and the third tones?

Solution. $\dfrac{3}{2} \times \dfrac{4}{3} = \dfrac{2}{1}$, so we get an octave.

Problem 217. A *minor sixth* complements a *major third* to form an octave; a *major sixth* complements a *minor third* to form an octave. What are the frequency ratios for minor and major sixths?

The same melody can be played in different keys; transposing it, we change the key. From the mathematical point of view transposition means that all frequencies are multiplied by a fixed number. So the frequency ratios remain unchanged, and consonant intervals remain consonant. You can observe this when you try your $33\frac{1}{3}$ record using a 45 rpm player. (The side effect is that music becomes not only higher in tone but also faster.)

Problem 218. How are the frequencies changed in this case?

Now we shall explain the connection between the well-tempered clavier and geometric progressions. It turns out that the following statement is true:

If the clavier (piano, harpsichord) is well tempered, that is, any melody can be transposed to start from any given tone, then the frequencies of the tones form a geometric progression.

Problem 219. Prove this statement.

Solution. Consider the *chromatic scale*, that is, the sequence of tones starting from a certain tone and going in increasing order (without gaps). Let's transpose it; we still get a chromatic scale. (If this new melody did not use some specific tone, then adding this tone would give us a melody that could not be transposed back.) If the initial tone of the transposed chromatic scale and the original scale are neighbors, then each tone is mapped to its neighbor tone after the transposition. In other words, we get the frequency of a neighbor tone when multiplying the frequency of the original tone by some constant. This is the definition of a geometric progression.

Now let us denote the frequency of tone A by a, and the common ratio of the tone progression by q. Then the chromatic scale starting with A has frequencies

$$a, \ aq, \ aq^2, \ aq^3, \ \ldots$$

This scale must include the A tone of the next octave, whose frequency is $2a$. So $2a = a \times q^n$, when n is the number of tones per octave in the chromatic scale. If you have access to a piano (or to a synthesizer,

if you cannot afford a piano or prefer "pop music"), you can easily find
that $n = 12$ (do not forget to count black keys). Therefore $q^{12} = 2$
and $q = \sqrt[12]{2}$ (see the section below about roots).

Now we can understand the inherent difficulty in tuning a piano:
the fifth (and other intervals, too) are not really true intervals. Indeed,
between the tones A and E there are 7 steps:

A	A♯ = B♭	B	C	C♯ = D♭	D	D♯ = E♭	E
a	aq	aq^2	aq^3	aq^4	aq^5	aq^6	aq^7

and to get a true fifth we need $q^7 = \dfrac{3}{2}$. But the requirements $q^{12} = 2$
and $q^7 = \dfrac{3}{2}$ are inconsistent; if both are fulfilled then

$$2^7 = \left(q^{12}\right)^7 = \left(q^7\right)^{12} = \left(\frac{3}{2}\right)^{12}$$

which is false. Using a pocket calculator we may find that when $q^{12} = 2$
we get $q^7 = 1.498307\ldots$, which is close but not equal to 1.5. For other
intervals the differences are even bigger:

Interval	should be	is about
		1.000000
		1.059463
		1.122462
minor third	1.2	1.189207
major third	1.25	1.259921
fourth	1.333...	1.334839
		1.414213
fifth	1.5	1.498307
minor sixth	1.6	1.587401
major sixth	1.666...	1.681792
		1.781797
		1.887748
octave	2.0	2.000000

In this table, the right column is a geometric progression accurate to
6 digits corresponding to a well-tempered clavier; the middle column
shows the "true" intervals, which are ratios of small integers.

Problem 220. We assumed as a given fact that an octave contains
12 tones and found that in this case the well-tempered clavier cannot

provide true fifths. What happens if we allow another number of tones in an octave? Is it possible to get true fifths or not?

Let us return to the history of music. In ancient times (before the eighteenth century), people tuned claviers (harpsichords at that time) trying to make at least some intervals harmonic (that is, corresponding to ratios of small integers). So the melodies sound nice in one key but become horrible when transposed into another key. Therefore some keys were avoided. A man named Andreas Werkmeister decided to go the other way and to make all intervals (that is, frequency ratios) the same. In this case, as we have seen, all intervals (except the octaves) are not exact but are close to the exact ones for all keys. It turns out that this is an acceptable solution. The great Bach honored this invention by writing his *Well-Tempered Clavier*. It contains two parts. Each part contains 24 preludes and fugues – one for each minor and major key.

Problem 221. Find a recording of Bach's *Well-Tempered Clavier* and enjoy it.

45 The sum of an infinite geometric progression

One of the famous "Paradoxes of Zeno" (Zeno was an ancient Greek philosopher) can be explained as follows. Assume that Achilles, who runs ten times faster than the turtle, starts to run after it. (The turtle runs away at the same time.) When Achilles comes to the place where the turtle was, it is not there but has moved on a distance equal to one tenth of the initial distance (between Achilles and the turtle). Achilles runs to that point – but at that time the turtle is again not there but has moved on a distance of one hundredth the initial distance, etc. This process has infinitely many stages; therefore Achilles will never meet the turtle. O.K?

We included this story in this section because the distances covered by Achilles form a geometric progression

$$1, \frac{1}{10}, \frac{1}{100}, \frac{1}{1000}, \cdots$$

whose common ratio is equal to $1/10$ (we assume that at the beginning the distance between Achilles and the turtle was equal to 1). So the total distance covered by Achilles is equal to the "sum of the infinite series"

$$1 + \frac{1}{10} + \frac{1}{100} + \frac{1}{1000} + \cdots$$

The pedantic view is that this infinite series has no sum (unless it is defined by a special definition) because when adding the numbers of an infinite series we never stop. And, of course, this is true. However, we shall not discuss this definition. Instead, we shall compute this undefined sum in different ways.

The first method is to denote this sum by S:

$$S = 1 + \frac{1}{10} + \frac{1}{100} + \frac{1}{1000} + \cdots$$

Then

$$10S = 10 + 1 + \frac{1}{10} + \frac{1}{100} + \cdots = 10 + S,$$

so

$$9S = 10, \quad S = \frac{10}{9}.$$

The second method is to add terms one by one:

$$1 + \frac{1}{10} = 1.1$$

$$1 + \frac{1}{10} + \frac{1}{100} = 1.11$$

$$1 + \frac{1}{10} + \frac{1}{100} + \frac{1}{1000} = 1.111$$

$$\cdots$$

After we add all terms, we get a periodic fraction $1.111\ldots$ equal to $1\frac{1}{9}$ (because $1/9 = 0.111\ldots$).

The third method is to apply the formula for the sum of the geometric progression:

$$1 + q + q^2 + q^3 + \cdots + q^{n-1} = \frac{q^n - 1}{q - 1}.$$

In our case $q = 1/10$ and n is infinitely big (so to speak). Then q^n is infinitely small (the bigger n is, the smaller $(1/10)^n$ is). Discarding it, we get the formula for the sum of an infinite geometric progression:

$$1 + q + q^2 + q^3 + \cdots = \frac{1}{1 - q}$$

(we have changed the signs of the numerator and the denominator). Recalling that $q = 1/10$, we get the answer $\dfrac{1}{0.9} = \dfrac{10}{9}$.

The fourth method. Let us return to Achilles and the turtle. Our common sense says that Achilles will meet the turtle after some distance S. During the race the turtle's speed is ten times less than that of Archilles, so the turtle covers the distance $S/10$. The initial distance (as we assume) was 1, so we get the equation

$$S - \frac{1}{10}S = 1.$$

Therefore, $(9/10)S = 1$ and $S = 10/9$.

Imagine now that Achilles is running ten times more slowly than the turtle. When he comes to the place where the turtle was, it is at the distance ten times further than the initial one. When Achilles comes to that place, the turtle is far away – at the distance that is one hundred times further than the initial one, etc. So we come to the sum

$$1 + 10 + 100 + \cdots$$

Of course, Achilles will never meet the turtle. But nevertheless we can substitute 10 for q in the formula

$$1 + q + q^2 + q^3 + \cdots = \frac{1}{1-q}.$$

and get an (absurd) answer

$$1 + 10 + 100 + 1000 + \cdots = \frac{1}{1-10} = -\frac{1}{9}.$$

Problem 222. Is it possible to give a reasonable interpretation of the (absurd) statement "Achilles will meet the turtle after running $-1/9$ meters"?

Hint. Yes, it is.

46 Equations

When we write, for instance, the equality

$$(a + b)^2 = a^2 + 2ab + b^2,$$

it has the following meaning: For any numbers a and b, the left-hand side and the right-hand side are equal. Such equalities are called identities. An identity may be proved (if we are lucky enough to transform the left-hand side to be equal to the right-hand side using algebraic rules). An identity may be refuted (if we managed to find values of variables such that the left-hand side is not equal to the right-hand side).

An equation also consists of a left-hand side and right-hand side connected by the equality sign, but the goal is different: it must be *solved*. To solve an equation means to find values of the variable(s) for which the left-hand side is equal to the right-hand side.

For example, the equation

$$5x + 3 = 2x + 7$$

may be solved as follows: Subtract $2x + 3$ from both sides; you get the equivalent equation

$$3x = 4$$

(the equivalence means that if one of the equations is true for some x then the other one is also true for this x). Now dividing both sides by 3 we get

$$x = \frac{4}{3}.$$

So we say: "the equation $5x + 3 = 2x + 7$ has the unique solution $x = 4/3$".

Remark. The equation

$$\frac{x + 1}{x + 2} = 1$$

has no solutions. (Proof: if $\dfrac{x + 1}{x + 2} = 1$ then $x + 1 = x + 2$, which is impossible.) However, mathematicians do not say that this equation is unsolvable. On the contrary, they say that the equation is solved after they proved that it has no solutions. So "to solve an equation" means to find all solutions or to prove that there are no solutions.

47 A short glossary

unknowns	letters used in an equation
to *solve* an equation	to find all values of unknowns such that the left-hand side is equal to the right-hand side; to find all solutions
a *solution* of an equation	a set of values for the unknown for which the left-hand side is equal to the right-hand side (sometimes solutions are called *roots* when speaking about an equation with only one unknown)
equivalent equations	equations having the same solutions; equations that are true or false simultaneously, for the same values of the unknowns

48 Quadratic equations

By a quadratic equation, we mean an equation of the form

$$ax^2 + bx + c = 0,$$

where a, b, c are some fixed numbers and x is an unknown.

Problem 223. Solve the quadratic equation $x^2 - 3x + 2 = 0$.

Solution. Factor the left-hand side: $x^2 - 3x + 2 = (x - 1)(x - 2)$. Therefore, the equation may be rewritten as $(x - 1)(x - 2) = 0$. This equality is true in two cases: either $x - 1 = 0$ (so $x = 1$) or $x - 2 = 0$ (so $x = 2$). Thus, this equation has two roots, $x = 1$ and $x = 2$.

Problem 224. Solve the equations:

(a) $x^2 - 4 = 0$;   (b) $x^2 + 2 = 0$;

(c) $x^2 - 2x + 1 = 0$;   (d) $x^2 - 2x + 1 = 9$;

(e) $x^2 - 2x - 8 = 0$;   (f) $x^2 - 2x - 3 = 0$;

(g) $x^2 - 5x + 6 = 0$;   (h) $x^2 - x - 2 = 0$.

If in the equation

$$ax^2 + bx + c = 0$$

the coefficient a is equal to zero then the equation takes the form

$$bx + c = 0$$

and has the unique solution

$$x = -\frac{c}{b}.$$

Problem 225. Strictly speaking, the last sentence is wrong; when $b = 0$ the quotient c/b in undefined. How are we to correct this error?

If in the equation

$$ax^2 + bx + c = 0$$

the coefficient a is nonzero then we may divide by a and get an equivalent equation

$$x^2 + \frac{b}{a}x + \frac{c}{a} = 0.$$

So if we are able to solve a *reduced* quadratic equation (where x^2 has a coefficient 1) we can solve any quadratic equation. Usually the reduced quadratic equation is written as

$$x^2 + px + q = 0.$$

49 The case $p = 0$. Square roots

Let us start with the equation $x^2 + q = 0$. Three cases are possible:

(a) $q = 0$. The equation $x^2 = 0$ has a unique solution $x = 0$.

(b) $q > 0$. The equation has no solutions because the nonnegative number x^2 added to a positive number q cannot be equal to 0.

(c) $q < 0$. The equation may be rewritten as $x^2 = -q$ and we have to look for numbers whose square is a (positive) number $-q$.

Fact. For any positive number c there is a positive number whose square is c. It is called the square root of c; its notation is $\sqrt{c}$.

We met with $\sqrt{2}$ in factoring $x^2 - 2 = (x - \sqrt{2})(x + \sqrt{2})$. Now we use $\sqrt{c}$ for a similar purpose.

How to solve the equation $x^2 = c$:

$$x^2 - c = 0;$$
$$x^2 - (\sqrt{c})^2 = 0;$$
$$(x - \sqrt{c})(x + \sqrt{c}) = 0;$$

the last equation has two solutions, $x = \sqrt{c}$ and $x = -\sqrt{c}$ (and no other solutions).

The reader may ask now, why are we considering this? When $x = \sqrt{c}$ then $x^2 = c$ by definition (and when $x = -\sqrt{c}$, too). Yes, this is true. But we have proved also that *there is no other solution* (because if $x \neq \pm\sqrt{c}$ then both factors are nonzero).

Now let us return to the fact claimed above, the existence of a square root. Assume that we start with $x = 0$ and then x increases gradually. Its square x^2 also increases (greater values of x correspond to greater values of x^2). At the beginning, $x^2 = 0$ and x^2 is less than c. When x is very big, x^2 is even bigger and therefore $x^2 > c$ for x big enough. So x^2 was *smaller* than c and becomes *greater* than c. Therefore, it must cross this boundary sometime – for some x the value of x^2 must be *equal* to c.

In the last sentence the word "therefore" stands for several chapters of a good calculus textbook, where the existence of such an x is proved, based on considerations of continuity.

These days, when square roots can be found on almost any calculator, it is almost impossible to imagine the shock caused by square roots for ancient Greeks. They found that the square root of 2 cannot be written as a quotient of two integers – and they did not know any other numbers, so it was a crash of their foundations.

Problem 226. Prove that $\sqrt{2} \neq \dfrac{m}{n}$ for any integer m and n. In other words, $\sqrt{2}$ is irrational (rational numbers are fractions with integers as numerator and denominator).

Solution. Assume that $\sqrt{2} = \dfrac{m}{n}$. Three cases are possible:

(a) both m and n are odd;

(b) m is even and n is odd;

(c) m is odd and n is even.

(The fourth case "m and n are even" may be ignored, because we could divide m and n by 2 several times until at least one of them would be odd and we would get one of the cases (a)–(c).)

Let us show that cases (a)–(c) are all impossible. Recall that any even number can be represented as $2k$ for some integer n and any odd number can be represented as $2k + 1$ for some integer k. So let us go through all three cases.

(a) Assume that $\sqrt{2} = \dfrac{2k + 1}{2l + 1}$; then

$$\left(\frac{2k + 1}{2l + 1}\right)^2 = 2,$$

$$\frac{(2k + 1)^2}{(2l + 1)^2} = 2,$$

$$(2k + 1)^2 = 2 \cdot (2l + 1)^2,$$

$$4k^2 + 4k + 1 = 2 \cdot (2l + 1)^2.$$

Contradiction: (even number) $+ 1 =$ (even number).

(b) Assume that $\sqrt{2} = \dfrac{2k}{2l + 1}$; then

$$\left(\frac{2k}{2l + 1}\right)^2 = 2,$$

$$(2k)^2 = 2 \cdot (2l + 1)^2,$$

$$4k^2 = 2 \cdot (4l^2 + 4l + 1),$$

$$2k^2 = 4l^2 + 4l + 1.$$

Contradiction: (even number) $=$ (even number) $+ 1$.

(c) Asssume that $\sqrt{2} = \dfrac{2k + 1}{2l}$; then

$$(2k + 1)^2 = 2 \cdot (2l)^2,$$

$$4k^2 + 4k + 1 = 2 \cdot (2l)^2.$$

Contradiction: (even number) $+ 1 =$ (even number).

So all three cases are impossible.

Problem 227. Prove that $\sqrt{3}$ is irrational.

Hint. Any integer has one of the forms $3k$, $3k+1$, $3k+2$.

When we claim that we have solved the equation $x^2 - 2 = 0$ and the answer is "$x = \sqrt{2}$ or $x = -\sqrt{2}$", we are in fact cheating. To tell the truth, we have not solved this equation but confessed our inability to solve it; $\sqrt{2}$ means nothing except "the positive solution of the equation $x^2 - 2 = 0$".

50 Rules for square roots

Problem 228. Prove that (for $a, b \geq 0$)

$$\sqrt{ab} = \sqrt{a} \cdot \sqrt{b}.$$

Solution. To show that $\sqrt{a} \cdot \sqrt{b}$ is the square root of ab we must (according to the definition of square roots) prove that it is a nonnegative number whose square is ab:

$$\left(\sqrt{a} \cdot \sqrt{b}\right)^2 = \left(\sqrt{a}\right)^2 \cdot \left(\sqrt{b}\right)^2 = a \cdot b.$$

Problem 229. Prove that for $a \geq 0$, $b > 0$

$$\sqrt{\frac{a}{b}} = \frac{\sqrt{a}}{\sqrt{b}}.$$

The following question is a traditional trap used by examiners to catch innocent pupils.

Problem 230. Is the equality $\sqrt{a^2} = a$ true for all a?

Solution. No. When a is negative, $\sqrt{a^2}$ is equal to $-a$. The correct statement is $\sqrt{a^2} = |a|$ where

$$|a| = \begin{cases} a, & \text{if } a \geq 0 \\ -a, & \text{if } a < 0 \end{cases}$$

Problem 231. Prove that

(a) $\dfrac{1}{2 + \sqrt{3}} = 2 - \sqrt{3}$;

(b) $\dfrac{1}{\sqrt{7} - \sqrt{5}} = \dfrac{\sqrt{5} + \sqrt{7}}{2}$.

Problem 232. Which is bigger: $\sqrt{1001} - \sqrt{1000}$, or $1/10$?

Problem 233. Simplify the expression $\sqrt{3 + 2\sqrt{2}}$

Solution. $3+2\sqrt{2} = 1+2+2\sqrt{2} = 1+(\sqrt{2})^2+2\sqrt{2} = (1+\sqrt{2})^2$. So we get the answer, $1+\sqrt{2}$.

Problem 234. Dan simplified an expression as follows:

$$\sqrt{3-2\sqrt{2}} = \sqrt{1+2-2\sqrt{2}} =$$
$$= \sqrt{1+(\sqrt{2})^2-2\sqrt{2}} = \sqrt{(1-\sqrt{2})^2} = 1-\sqrt{2}.$$

Do you approve of his simplification?

Solution. The correct answer is $\sqrt{2}-1$ because $1-\sqrt{2} < 0$.

51 The equation $x^2 + px + q = 0$

Problem 235. Solve the equation

$$x^2 + 2x - 6 = 0.$$

Solution. The equation $x^2 + 2x - 6 = 0$ may be rewritten as follows:

$$(x^2 + 2x + 1) - 7 = 0;$$
$$(x+1)^2 - 7 = 0;$$
$$(x+1)^2 = 7;$$
$$x+1 = \sqrt{7} \quad \text{or} \quad x+1 = -\sqrt{7};$$
$$x = -1+\sqrt{7} \quad \text{or} \quad x = -1-\sqrt{7}.$$

The same method can be applied to other equations.

Problem 236. Solve the equation

$$x^2 + 2x - 8 = 0.$$

Problem 237. Solve the equation

$$x^2 + 3x + 1 = 0.$$

Solution. Transform the left-hand side:

$$x^2 + 3x + 1 = x^2 + 2 \cdot \frac{3}{2}x + \left(\frac{3}{2}\right)^2 - \left(\frac{3}{2}\right)^2 + 1 =$$
$$= \left(x+\frac{3}{2}\right)^2 - \frac{9}{4} + 1 = \left(x+\frac{3}{2}\right)^2 - \frac{5}{4}.$$

Now the equation can be written as follows:

$$\left(x + \frac{3}{2}\right)^2 = \frac{5}{4},$$

$$x + \frac{3}{2} = \sqrt{\frac{5}{4}} \quad \text{or} \quad x + \frac{3}{2} = -\sqrt{\frac{5}{4}},$$

$$x = -\frac{3}{2} + \sqrt{\frac{5}{4}} \quad \text{or} \quad x = -\frac{3}{2} - \sqrt{\frac{5}{4}}.$$

Remark. The answer to the preceding problem is usually written as

$$x = -\frac{3}{2} \pm \sqrt{\frac{5}{4}}.$$

Problem 238. Solve the equation $x^2 - 2x + 2 = 0$.

Solution. $x^2 - 2x + 2 = (x^2 - 2x + 1) + 1 = (x - 1)^2 + 1$. The equation $(x - 1)^2 + 1 = 0$ has no roots because its left-hand side is never less than 1 (a square is always nonnegative).

The method shown above is called "completing the square". In the general case it looks as follows:

$$x^2 + px + q = 0$$

$$\left(x^2 + 2 \cdot \frac{p}{2} \cdot x + \left(\frac{p}{2}\right)^2\right) - \left(\frac{p}{2}\right)^2 + q = 0$$

$$\left(x + \frac{p}{2}\right)^2 = \left(\frac{p}{2}\right)^2 - q = \frac{p^2}{4} - q$$

Now three cases are possible:

- If $\dfrac{p^2}{4} - q > 0$ then two solutions exist:

$$x + \frac{p}{2} = \pm\sqrt{\frac{p^2}{4} - q}.$$

  Thus,

$$x = -\frac{p}{2} \pm \sqrt{\frac{p^2}{4} - q}.$$

- If $\dfrac{p^2}{4} - q = 0$ then there is one solution:

$$x = -\frac{p}{2}.$$

- If $\dfrac{p^2}{4} - q < 0$ then there are no solutions.

Often all three cases are included in a single formula:

$$x_{1,2} = -\frac{p}{2} \pm \sqrt{\frac{p^2}{4} - q}$$

and people say that when $\dfrac{p^2}{4} - q = 0$ the solutions x_1 and x_2 coincide (because the square root of 0 is 0) and when $\dfrac{p^2}{4} - q < 0$ this formula gives no solutions, because the square root of a negative number is undefined. (To tell you the whole truth, in the latter case mathematicians agree that the square root of a negative number exists but is imaginary and there are two so-called complex roots. But this is another topic.)

We see that the sign of $D = \dfrac{p^2}{4} - q$ plays a crucial role (it determines how many solutions the equation has).

52 Vieta's theorem

Theorem. If a quadratic equation $x^2 + px + q$ has two (different) roots α and β then

$$\begin{aligned} \alpha + \beta &= -p \\ \alpha \cdot \beta &= q. \end{aligned}$$

Corollary. If a quadratic equation $x^2 + px + q$ has two different roots α and β then

$$x^2 + px + q = (x - \alpha)(x - \beta).$$

This is another form of the same assertion because

$$(x - \alpha)(x - \beta) = x^2 - (\alpha + \beta)x + \alpha\beta$$

and two polynomials are equal if they have equal coefficients.

First proof. According to the formula for roots we have

$$\alpha = -\frac{p}{2} - \sqrt{D}, \quad \beta = -\frac{p}{2} + \sqrt{D}$$

where $D = \dfrac{p^2}{4} - q$. (Or vice versa:

$$\alpha = -\frac{p}{2} + \sqrt{D}, \quad \beta = -\frac{p}{2} - \sqrt{D},$$

but this makes no difference.) Then

$$\alpha + \beta = -\frac{p}{2} - \sqrt{D} - \frac{p}{2} + \sqrt{D} = -p$$

and

$$\alpha\beta = \left(-\frac{p}{2} - \sqrt{D}\right)\left(-\frac{p}{2} + \sqrt{D}\right) = \left(\frac{p}{2}\right)^2 - \left(\sqrt{D}\right)^2 =$$

$$= \frac{p^2}{4} - D = \frac{p^2}{4} - \frac{p^2}{4} + q = q.$$

That's what we want.

Second proof. Let us try to prove Vieta's theorem in the form stated in the corollary. We know that if a polynomial $P(x)$ has different roots α and β then it can be factored:

$$P(x) = (x - \alpha)(x - \beta)R(x)$$

where $R(x)$ is some polynomial. In our case (when P has degree 2) the polynomial R must be a constant (otherwise the degree of the right-hand side would be too big), and this constant is equal to 1, because the x^2-coefficients in $x^2 + px + q$ and $(x - \alpha)(x - \beta)$ are the same. Therefore

$$x^2 + px + q = (x - \alpha)(x - \beta).$$

The theorem is proved.

Problem 239. Can you generalize Vieta's theorem to the case of a quadratic equation having only one root? Are both proofs still valid for this case?

Problem 240. (Vieta's theorem for a cubic equation) Assume that a cubic equation $x^3 + px^2 + qx + r = 0$ has three different roots α, β, γ. Prove that

$$\begin{aligned}
\alpha + \beta + \gamma &= -p \\
\alpha\beta + \alpha\gamma + \beta\gamma &= q \\
\alpha\beta\gamma &= -r
\end{aligned}$$

Problem 241. The equation $x^2 + px + q = 0$ has roots x_1 and x_2. Find $x_1^2 + x_2^2$ (as an expression containing p and q).

Solution. $x_1^2 + x_2^2 = x_1^2 + 2x_1x_2 + x_2^2 - 2x_1x_2 = (x_1 + x_2)^2 - 2x_1x_2 = p^2 - 2q$.

Problem 242. The equation $x^2 + px + q = 0$ has roots x_1 and x_2. Find $(x_1 - x_2)^2$ (as an expression containing p and q).

Solution. $(x_1 - x_2)^2 = x_1^2 - 2x_1x_2 + x_2^2 = x_1^2 + 2x_1x_2 + x_2^2 - 4x_1x_2 = (x_1 + x_2)^2 - 4x_1x_2 = p^2 - 4q$.

Another solution. $x_1 - x_2$ is the difference between the roots; looking at the formula for the roots, we see that it is equal to $2\sqrt{D}$, so $(x_1 - x_2)^2 = 4D = 4(\frac{p^2}{4} - q) = p^2 - 4q$.

Problem 243. A cubic equation $x^3 + px^2 + qx + r = 0$ has three different roots x_1, x_2, x_3. Find

$$(x_1 - x_2)^2(x_2 - x_3)^2(x_1 - x_3)^2$$

as an expression containing p, q, r. This polynomial in p, q, r is called the *discriminant* of the cubic equation. As in the case of a quadratic equation (see page 107), it is small when two roots are close to each other.

Problem 244. The equation $x^2 + px + q = 0$ has roots x_1, x_2; the equation $y^2 + ry + s = 0$ has roots y_1, y_2. Find

$$(y_1 - x_1)(y_2 - x_1)(y_1 - x_2)(y_2 - x_2)$$

as a polynomial of p, q, r, s. (This polynomial is called the resultant of two quadratic polynomials; it is equal to zero if these two polynomials have a common root.)

Vieta's theorem allows us to construct a quadratic equation with given roots. More precisely, we should not say "Vieta's theorem" but "the converse to Vieta's theorem"; here it is:

Theorem. If α and β are any numbers, $p = -(\alpha + \beta)$, $q = \alpha\beta$, then the equation $x^2 + px + q = 0$ has roots α and β.

The proof is trivial: The equation $(x - \alpha)(x - \beta) = 0$ evidently has roots α and β. Multiplying the terms in parentheses we see that it is the equation $x^2 + px + q = 0$.

Problem 245. Find a quadratic equation with integer coefficients having $4 - \sqrt{7}$ as one of the roots.

Hint. The second root is $4 + \sqrt{7}$.

Problem 246. The integers p, q are coefficients of the quadratic equation $x^2 + px + q = 0$, which has two roots. Prove that

(a) the sum of squares of its roots is an integer;

(b) the sum of cubes of its roots is an integer;

(c) the sum of nth powers of its roots is an integer (for any natural number n)

Problem 247. (a) Prove that the square of any number of the form $a + b\sqrt{2}$ (where a, b are integers) also has this form (that is, is equal to $k + l\sqrt{2}$ for some integer k, l).

(b) Prove the same for $(a + b\sqrt{2})^n$ for any integer $n > 1$.

(c) The number $(a + b\sqrt{2})^n$ is equal to $k + l\sqrt{2}$ (here a, b, k, l are integers). What can be said about $(a - b\sqrt{2})^n$?

(d) Prove that there are infinitely many integers a, b such that $a^2 - 2b^2 = 1$.

Solution of (d). Let us start from the solution $3^2 - 2 \cdot 2^2 = 1$, and rewrite this equality as $(3 + 2\sqrt{2})(3 - 2\sqrt{2}) = 1$. Consider the nth powers of both sides: $(3 + 2\sqrt{2})^n(3 - 2\sqrt{2})^n = 1$. The number $(3 + 2\sqrt{2})^n$ is equal to $k + l\sqrt{2}$ for some integers k, l. Thus $(3 - 2\sqrt{2})^n$ is equal to $k - l\sqrt{2}$ and we get the equality

$$(k + l\sqrt{2})(k - l\sqrt{2}) = k^2 - 2l^2 = 1.$$

Therefore, k, l satisfy the equation.

For example, $(3 + 2\sqrt{2})^2 = 9 + 8 + 12\sqrt{2} = 17 + 12\sqrt{2}$. So $17, 12$ must satisfy the equation. Is this true? $17^2 - 2 \cdot 12^2 = 289 - 2 \cdot 144 = 289 - 288 = 1$. Our theory works!

Problem 248. Prove that the equation $x^2 + px + q = 0$ has two solutions having different signs if and only if $q < 0$

Solution. If the roots have opposite signs, then (recall Vieta's theorem) the coefficient q, being equal to their product, is negative. In the opposite direction, if the product of two roots is negative, they have opposite sign. (But we must be sure that the roots do exist; to check this, we look at $D = \frac{p^2}{4} - q$; if $q < 0$, then $D > 0$.)

Another explanation can be given as follows. Assume that $q < 0$. Then the value of the expression $x^2 + px + q$ is negative when $x = 0$. When x increases and becomes very big, $x^2 + px + q$ becomes positive (x^2 "outweighs" $px + q$). So $x^2 + px + q$ must cross the zero boundary somewhere in between – and the equation has a positive root. A similar argument shows that it also has a negative root.

53 Factoring $ax^2 + bx + c$

Problem 249. Factor $2x^2 + 5x - 3$.

Solution. Taking the factor 2 out of the parentheses, we get

$$2x^2 + 5x - 3 = 2\left(x^2 + \frac{5}{2}x - \frac{3}{2}\right).$$

Solving the equation $x^2 + \frac{5}{2}x - \frac{3}{2} = 0$ we get

$$x_{1,2} = -\frac{5}{4} \pm \sqrt{\frac{25}{16} + \frac{3}{2}} = -\frac{5}{4} \pm \sqrt{\frac{49}{16}} = -\frac{5}{4} \pm \frac{7}{4}$$

so $x_1 = -3$, $x_2 = \frac{1}{2}$. According to the corollary of Vieta's theorem, we get

$$x^2 + \frac{5}{2}x - \frac{3}{2} = (x - (-3))\left(x - \frac{1}{2}\right) = (x + 3)\left(x - \frac{1}{2}\right)$$

and

$$2x^2 + 5x - 3 = (x + 3)(2x - 1).$$

Problem 250. Factor $2x^2 + 2x + \frac{1}{2}$.

Problem 251. Factor $2a^2 + 5ab - 3b^2$.

Solution.

$$2a^2 + 5ab - 3b^2 = b^2\left(2\frac{a^2}{b^2} + 5\frac{a}{b} - 3\right).$$

Denote $\frac{a}{b}$ by x and use the factorization $2x^2 + 5x - 3 = (x+3)(2x-1)$. Then you can continue the equality:

$$\cdots = b^2\left(\frac{a}{b} + 3\right)\left(2\frac{a}{b} - 1\right) = (a + 3b)(2a - b).$$

54 A formula for $ax^2 + bx + c = 0$ (where $a \neq 0$)

Dividing by a, we get

$$x^2 + \frac{b}{a}x + \frac{c}{a} = 0$$

and we can apply the formula for the equation

$$x^2 + px + q = 0$$

with $p = \dfrac{b}{a}$, $q = \dfrac{c}{a}$. We get

$$x_{1,2} = -\frac{b}{2a} \pm \sqrt{\left(\frac{b}{2a}\right)^2 - 4\frac{c}{a}} = -\frac{b}{2a} \pm \sqrt{\frac{b^2 - 4ac}{4a^2}} =$$

$$= -\frac{b}{2a} \pm \frac{\sqrt{b^2 - 4ac}}{2a} = \frac{-b \pm \sqrt{b^2 - 4ac}}{2a}.$$

The expression $D = b^2 - 4ac$ is called the *discriminant* of the equation $ax^2 + bx + c = 0$. If it is positive, the equation has two roots. If $D = 0$ the equation has one root. If $D < 0$ the equation has no roots.

Problem 252. We replaced

$$\sqrt{\frac{b^2 - 4ac}{4a^2}} = \frac{\sqrt{b^2 - 4ac}}{\sqrt{4a^2}}$$

by

$$\frac{\sqrt{b^2 - 4ac}}{2a}$$

but as we mentioned above, $\sqrt{4a^2}$ is equal not to $2a$ but to $|2a|$. Why does it not matter here?

Problem 253. Assume that the equation $ax^2 + bx + c = 0$ has roots x_1 and x_2. What are the roots of the equation $cx^2 + bx + a = 0$?

Solution. If $ax^2 + bx + c = 0$ then (divide by x^2)

$$a + \frac{b}{x} + \frac{c}{x^2} = 0, \text{ that is, } c \cdot \left(\frac{1}{x}\right)^2 + b \cdot \left(\frac{1}{x}\right) + a = 0.$$

So $\dfrac{1}{x_1}$ and $\dfrac{1}{x_2}$ will be the roots of the equation $cx^2 + bx + a = 0$.

Remark. We assumed implicitly that $x_1 \neq 0$, $x_2 \neq 0$. If one of the roots x_1 and x_2 is equal to 0 then (according to Vieta's theorem) c is equal to 0 and the equation $cx^2 + bx + a = 0$ has at most one root.

55 One more formula concerning quadratic equations

The formula
$$x_{1,2} = \frac{-b \pm \sqrt{b^2 - 4ac}}{2a}$$
is well known to millions of school pupils all over the world. But there is another formula that has an equal right to be studied but is less known. Here it is:
$$x_{1,2} = \frac{2c}{-b \pm \sqrt{b^2 - 4ac}}$$

Let us prove it. If x is a root of the equation $ax^2 + bx + c = 0$ then $y = 1/x$ is a root of the equation $cy^2 + by + a = 0$, therefore
$$y_{1,2} = \frac{-b \pm \sqrt{b^2 - 4ac}}{2c}$$
and
$$x_{1,2} = \frac{1}{y_{1,2}} = \frac{2c}{-b \pm \sqrt{b^2 - 4ac}}.$$

Problem 254. Check by a direct calculation that
$$\frac{-b \pm \sqrt{b^2 - 4ac}}{2a} = \frac{2c}{-b \mp \sqrt{b^2 - 4ac}}.$$

(we used $\mp$ and $\pm$ having in mind that plus in the left-hand side corresponds to the minus in the right-hand side and vice versa).

56 A quadratic equation becomes linear

Look at the quadratic equation $ax^2 - x + 1 = 0$. According to our general rule, it has two roots if and only if its discriminant $D = 1 - 4a$ is positive, that is, when $a < 1/4$.

Problem 255. Is this true?

Solution. No; when $a = 0$ the equation is not a quadratic one, it becomes $-x + 1 = 0$ and has only one root $x = 1$.

A pedant will describe what happens saying "our general rule is not applicable, because the equation is not quadratic". And he is right. But how can it be? We had an equation with two roots and were changing the coefficient. Suddenly one root disappeared, when a became zero. What happened to it?

To answer this question, let us look at the second formula for the roots of a quadratic equation:

$$x_{1,2} = \frac{2}{1 \pm \sqrt{1 - 4a}}.$$

If a is close to zero then $\sqrt{1 - 4a} \approx 1$, so

$$x_1 = \frac{2}{1 + \sqrt{1 - 4a}} \approx \frac{2}{1 + 1} = 1,$$

but

$$x_2 = \frac{2}{1 - \sqrt{1 - 4a}} = \frac{2}{\text{number close to zero}},$$

that is, x_2 is very big. So while a tends to zero, the root x_1 tends to 1 and x_2 goes to infinity (and returns from the other side of infinity).

One can see in detail how this happens looking at the following picture:

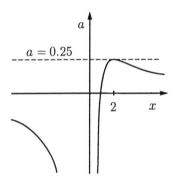

This picture shows points $\langle x, a \rangle$ such that $ax^2 - x + 1 = 0$. In other words, it shows the graph of the function $a = (x - 1)/x^2$. To find the solutions of the equation $ax^2 - x + 1 = 0$ for a given a on the picture, we must intersect a horizontal straight line having height a with our graph. Assume that this horizontal line is moving downwards. At the beginning (when $a > 0.25$) it has no intersections (and the equation has no solutions). When $a = 0.25$ there is one intersection point, which splits immediately into two points when a becomes less than 0.25. One of the points is moving left, the other is moving right. The point moving right goes to plus infinity when a tends to zero, then disappears (when $a = 0$) and then returns from minus infinity. Then (when a becomes more and more negative), both roots go to zero from opposite sides.

Problem 256. What happens with the roots of equations
(a) $x^2 - x - a = 0$;   (b) $x^2 - ax + 1 = 0$
as a changes?

57 The graph of the quadratic polynomial

The graph of $y = x^2$ looks as follows:

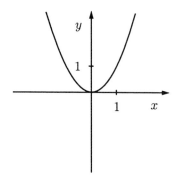

Using this graph we may draw graphs of other polynomials of degree 2. The graph of $y = ax^2$ (where a is a constant) can be obtained from $y = x^2$ by stretching (when $a > 1$) or shrinking (when $0 < a < 1$) in a vertical direction:

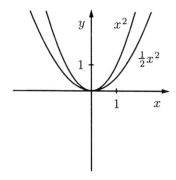

When a is negative the graph is turned upside down:

110

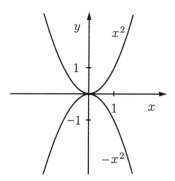

The graph $y = x^2 + c$ can be obtained from the graph of $y = x^2$ by a vertical translation by c (up if $c > 0$, down if $c < 0$).

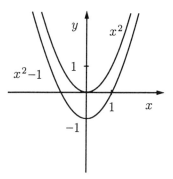

In the same way, $y = ax^2 + c$ can be obtained from $y = ax^2$.

It is more difficult to understand what corresponds to a horizontal translation of the graph. Let us consider an example and compare the graphs $y = \frac{1}{2}x^2$ and $y = \frac{1}{2}(x + 1)^2$. Let us start with one specific value of x; assume that $x = -3$. For this x the expression $\frac{1}{2}(x + 1)^2$ is equal to $\frac{1}{2}(-2)^2$, that is, has the same value as $\frac{1}{2}x^2$ when $x = -2$. In general, the value of $\frac{1}{2}(x + 1)^2$ for any value x coincides with the value of $\frac{1}{2}x^2$ for some other value x (greater by 1).

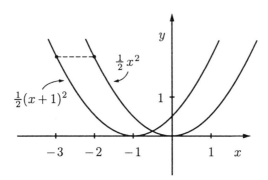

In terms of our graph this means that any point of the graph $y = \frac{1}{2}(x+1)^2$, when moved 1 unit to the right, becomes a point of the graph $y = \frac{1}{2}x^2$. Therefore, we get the graph $y = \frac{1}{2}x^2$ by translating the graph $y = \frac{1}{2}(x+1)^2$ one unit to the right, and vice versa, we get the graph $y = \frac{1}{2}(x+1)^2$ by translating the graph $y = \frac{1}{2}x^2$ one unit to the left.

The general rule is as follows: a graph $y = a(x+m)^2$ can be obtained from the graph of $y = ax^2$ by an m-unit shift to the left (when $m > 0$; when $m < 0$ we use a right shift).

Now we can get any graphs of the form

$$y = a(x+m)^2 + n$$

from the graph of $y = x^2$ in three stages:

(a) Stretch it vertically a times and you get $y = ax^2$.

(b) Move it m units to the left and you get $y = a(x+m)^2$.

(c) Move it n units up and you get $y = a(x+m)^2 + n$.

Problem 257. Find the coordinates of the top point (or the bottom point – it depends on the sign of a) of a graph $y = a(x+m)^2 + n$.

Answer. Its coordinates are $\langle -m, n \rangle$.

Problem 258. Is the ordering of operations (a), (b), and (c) important? Do we get the same graph applying, for example, (c), then (b), and then (a) to the graph $y = x^2$?

112

Answer. The ordering of operations is important. We get $x^2 + n$ after (**c**), then $(x+m)^2 + n$ after (**b**) and finally $a(x+m)^2 + an$ after (**a**). So we get an instead of n.

Problem 259. There are six possible orderings of operations (**a**), (**b**), and (**c**). Do we get six different graphs or do some of the graphs coincide?

Now we are able to draw the graph of any quadratic polynomial, because any quadratic polynomial may be converted to the form $a(x+m)^2 + n$ by completing the square (as we did for the formula for the roots):

$$ax^2 + bx + c = a\left(x^2 + \frac{b}{a}x\right) + c =$$

$$= a\left(x^2 + 2 \cdot \frac{b}{2a} \cdot x + \left(\frac{b}{2a}\right)^2 - \left(\frac{b}{2a}\right)^2\right) + c =$$

$$= a\left(x + \frac{b}{2a}\right)^2 - \frac{b^2}{4a} + c.$$

Denote $\dfrac{b}{2a}$ by m and $-\dfrac{b^2}{4a} + c$ by n and you get the desired result.

Problem 260. How can you determine the signs of a, b, c by looking at the graph of $y = ax^2 + bx + c$?

Answer. If water can be kept in this graph then $a > 0$, otherwise $a < 0$. The sign of b/a is determined by the x-coordinate of the vertex of the graph (the left half of the plane corresponds to positive b/a). The sign of c can be found by looking at the intersection of the graph and y-axis (because $ax^2 + bx + c = c$ when $x = 0$).

Remark. Another rule for finding the sign of b: If the graph intersects the y-axis going upwards, then b is positive; if the graph intersects it going downwards, then b is negative. This rule can be explained by means of calculus. When the function $f(x) = ax^2 + bx + c$ is increasing near $x = 0$, its derivative $f'(x) = 2ax + b$ (which is equal to b when $x = 0$) is positive.

58 Quadratic inequalities

Problem 261. Solve the inequality $x^2 - 3x + 2 < 0$. (To "solve an inequality" means to find all values of the variables for which it is true.)

Solution. Factor the left-hand side:

$$x^2 - 3x + 2 = (x - 1)(x - 2)$$

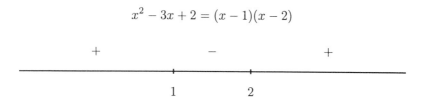

The left-hand side is zero when $x = 1$ or $x = 2$. When $x > 2$, both factors are positive (and the product is positive). When we go through the point $x = 2$ into the interval $(1, 2)$, the second factor becomes negative (and the product is negative). When we go through the point $x = 1$, both factors become negative and the product is positive again. Therefore, we get an answer that the inequality is true for $1 < x < 2$.

You can get the same answer looking at the graph $y = x^2 - 3x + 2$ ($x = 1$ and $x = 2$ are intersection points with x-axis).

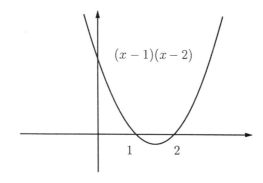

59 Maximum and minimum values of a quadratic polynomial

Problem 262. The sum of two numbers is equal to 1. What is the maximal possible value of its product?

Solution. Denote one of the numbers by x. Then the second number is $1 - x$, and their product is $x \cdot (1 - x) = x - x^2$. The graph of the quadratic polynomial $-x^2 + x$ is turned down and its roots are $x = 0$ and $x = 1$. Therefore its vertex, being in the middle, has $x = \frac{1}{2}$. Its value for $x = \frac{1}{2}$ (its maximal value) is $\frac{1}{2} \cdot (1 - \frac{1}{2}) = \frac{1}{4}$. So we get the answer that the maximal value is $\frac{1}{4}$.

Another solution. Assume that one of the numbers is $\frac{1}{2} + x$. Then the other number is $\frac{1}{2} - x$ and their product is

$$\left(\frac{1}{2} + x\right)\left(\frac{1}{2} - x\right) = \frac{1}{4} - x^2,$$

so the maximal value is obtained when $x = 0$ (and both numbers are equal to $\frac{1}{2}$).

Problem 263. Prove that a square has the maximum area of all rectangles having the same perimeter.

Problem 264. Prove that a square has the minimum perimeter of all rectangles having the same area.

Hint. Use the result of the preceding problem.

Problem 265. Find the minimal value of the expression $x + \dfrac{2}{x}$ for positive x.

Solution. Let us see what numbers $c > 0$ may be values of the expression $x + \dfrac{2}{x}$. In other words we want to know for which c the equation

$$x + \frac{2}{x} = c$$

has solutions. We may multiply this equation by x and ask for which c the resulting equation

$$x^2 + 2 = cx$$

has nonzero solutions. But no solutions of this equation are equal to zero ($x = 0$ is not a solution, $0^2 + 2 \neq c \cdot 0$). Therefore, the word "nonzero" may be omitted.

The equation $x^2 + 2 = cx$ may be rewritten as $x^2 - cx + 2 = 0$. It has solutions if and only if its discriminant

$$D = \left(\frac{c}{2}\right)^2 - 2$$

is nonnegative, that is, when $\left(\frac{c}{2}\right)^2 \geq 2$. The latter condition is satisfied when

$$\frac{c}{2} \geq \sqrt{2} \ \text{ or } \ \frac{c}{2} \leq -\sqrt{2}.$$

So the equation $x + \frac{2}{c} = c$ has solutions when $c \geq 2\sqrt{2}$ or $c \leq -2\sqrt{2}$.

Therefore, the minimum value of $x + \frac{2}{x}$ for positive x is $2\sqrt{2}$.

Another solution. The numbers x and $\frac{2}{x}$ may be considered as edges of a rectangle having area 2, and $x + \frac{2}{x}$ is its semiperimeter. It will be minimal when the rectangle is a square (see the preceding problem), that is, when $x = \frac{2}{x}$, $x^2 = 2$, $x = \sqrt{2}$. For such an x the value of $x + \frac{2}{x}$ is $2\sqrt{2}$.

60 Biquadratic equations

Problem 266. Solve the equation $x^4 - 3x^2 + 2 = 0$.

Solution. If x is a root of this equation, then $y = x^2$ is a root of the equation $y^2 - 3y + 2 = 0$, and vice versa. This quadratic equation (where y is considered an unknown) has roots

$$y_{1,2} = \frac{3 \pm \sqrt{9 - 8}}{2} = \frac{3 \pm 1}{2};$$

hence, $y_1 = 1$, $y_2 = 2$.

Therefore the solutions of the initial equation are all x such that $x^2 = 1$ or $x^2 = 2$. So it has four solutions:

$$x = 1, \ x = -1, \ x = \sqrt{2}, \ x = -\sqrt{2}.$$

The same method can be applied to any equation of the form

$$ax^4 + bx^2 + c = 0$$

(such equations are called biquadratic)

Problem 267. Construct a biquadratic equation

(a) having no solution;

(b) having exactly one solution;

(c) having exactly two solutions;

(d) having exactly three solutions;

(e) having exactly four solutions;

(f) having exactly five solutions.

Hint. One of the cases (a)–(f) is impossible.

Problem 268. What is the possible number of solutions of the equation

$$ax^6 + bx^3 + c = 0\,?$$

Hint. Remember that a, b or c may be equal to 0.

Answer. 0, 1, 2, or infinitely many.

Problem 269. The same question for the equation

$$ax^8 + bx^4 + c = 0.$$

61 Symmetric equations

Problem 270. Solve the equation

$$2x^4 + 7x^3 + 4x^2 + 7x + 2 = 0.$$

Solution. First of all, $x = 0$ is not a solution of this equation. Therefore we lose nothing dividing by x:

$$2x^2 + 7x + 4 + \frac{7}{x} + \frac{2}{x^2} = 0.$$

Now we group terms with equal coefficients and opposite powers of x:

$$2\left(x^2 + \frac{1}{x^2}\right) + 7\left(x + \frac{1}{x}\right) + 4 = 0.$$

Now we use that $x^2 + \dfrac{1}{x^2}$ may be expressed in terms of $x + \dfrac{1}{x}$:

$$\left(x + \frac{1}{x}\right)^2 = x^2 + 2 \cdot x \cdot \frac{1}{x} + \left(\frac{1}{x}\right)^2 = x^2 + \frac{1}{x^2} + 2,$$

and

$$x^2 + \frac{1}{x^2} = \left(x + \frac{1}{x}\right)^2 - 2.$$

Therefore, if x is a solution of the given equation, then $y = x + \dfrac{1}{x}$ is a solution of the equation $2(y^2 - 2) + 7y + 4 = 0$, or $2y^2 - 4 + 7y + 4 = 0$, or $2y^2 + 7y = 0$, or $y(2y + 7) = 0$, whose solutions are $y = 0$ and $y = -7/2$. Therefore the solutions of the initial equation are all x such that
$$x + \frac{1}{x} = 0 \quad \text{or} \quad x + \frac{1}{x} = -\frac{7}{2}.$$

Let us solve these two equations. The first one:
$$x + \frac{1}{x} = 0 \implies x^2 + 1 = 0 \implies x^2 = -1. \text{ No solutions.}$$

The second equation: $x + \dfrac{1}{x} = -\dfrac{7}{2}$ means (we know that $x \neq 0$) that $x^2 + 1 = -\dfrac{7}{2}x$, or $x^2 + \dfrac{7}{2}x + 1 = 0$; the roots are

$$x_{1,2} = \frac{-\dfrac{7}{2} \pm \sqrt{\dfrac{49}{4} - 4}}{2} = \frac{-\dfrac{7}{2} \pm \dfrac{\sqrt{33}}{2}}{2}$$

Answer. The given equation has two solutions:

$$x_1 = \frac{-7 - \sqrt{33}}{4}, \quad x_2 = \frac{-7 + \sqrt{33}}{4}.$$

62 How to confuse students on an exam

As usual, there are many ways to make evil use of knowledge. Here are the instructions for one of them, namely, how to invent a practically unsolvable equation.

1. Take a quadratic equation – preferably with non-integer roots, for example,
$$3x^2 + 2x - 10 = 0.$$
whose roots are $x_{1,2} = \dfrac{-2 \pm \sqrt{4 + 120}}{6} = \dfrac{-2 \pm \sqrt{124}}{6} = \dfrac{-1 \pm \sqrt{31}}{3}.$

2. Substitute some polynomial of degree 2 instead of x, for example, take $x = y^2 + y - 1$. You get

$$x^2 = (y^2 + y - 1)(y^2 + y - 1) = y^4 + 2y^3 - y^2 - 2y + 1,$$

$$3x^2 + 2x - 10 = \begin{array}{llll} 3y^4 & + 6y^3 & - 3y^2 & - 6y & + 3 \\ & & + 2y^2 & + 2y & - 2 \\ & & & & -10 \end{array}$$

$$= \quad 3y^4 \quad + 6y^3 \quad - y^2 \quad - 4y \quad - 9.$$

3. Ask the students to solve the equation

$$3y^4 + 6y^3 - y^2 - 4y - 9 = 0.$$

4. Wait 10 to 15 minutes.

5. Tell the students that their time is up and they failed.

6. If somebody complains that the problem is too difficult and could not be solved by standard methods, you can explain that in fact this equation can be easily reduced to a quadratic:

$$\begin{array}{lllll} & 3y^4 & + 6\,y^3 & - y^2 & - 4y & - 9 & = \\ = & 3y^4 & + 3\,y^3 & - 3y^2 & & \\ & & + 3\,y^3 & + 3y^2 & - 3y & \\ & & & - y^2 & - y & + 1 \\ & & & & & - 10 & = \end{array}$$

$$= 3y^2(y^2 + y - 1) + 3y(y^2 + y - 1) - (y^2 + y - 1) - 10 =$$

$$= 3(y^2 + y)(y^2 + y - 1) - (y^2 + y - 1) - 10.$$

If now we denote $y^2 + y - 1$ by x we get an equation

$$3(x + 1)x - x - 10 = 0,$$

$$3x^2 + 3x - x - 10 = 0,$$

$$x_{1,2} = \frac{-1 \pm \sqrt{31}}{3}$$

and it remains to solve the two equations

$$y^2 + y - 1 = \frac{-1 - \sqrt{31}}{3} \quad \text{and} \quad y^2 + y - 1 = \frac{-1 - \sqrt{31}}{3}.$$

That's all, isn't it?

Another efficient method is to choose two quadratic equations with non-integer roots, for example,

$$x^2 + x - 3 = 0 \quad \text{and} \quad x^2 + 2x - 1 = 0$$

and multiply them:

$$(x^2 + x - 3)(x^2 + 2x - 1) = x^4 + 3x^3 - 2x^2 - 7x + 3 = 0.$$

The resulting equation can be given to students without a big risk of seeing it solved. But don't lose the sheet of paper with the factoring; otherwise you will be caught by your own trap when somebody asks you to show the solution!

63 Roots

A square root of a is defined as a number whose square is equal to a. (To be exact, a square root of a *nonnegative* number a is a *nonnegative* number whose square is equal to a.) In the same way we can define other roots: a *cube root* of $A \geq 0$ is a number $x \geq 0$ such that $x^3 = a$, a *fourth root* of $a \geq 0$ is a number $x \geq 0$ such that $x^4 = a$, etc. The notation for the nth root of a is $\sqrt[n]{a}$.

Definition. An nth root of a nonnegative number a is a nonnegative number x such that $x^n = a$. (We assume that n is a positive integer.)

This definition raises several questions.

Question. What happens if there are many numbers x having this property?

Answer. This cannot happen. The greater a nonnegative number x, the greater is x^n (if in a product of nonnegative factors all factors increase, the product increases also). So different nonnegative values of x have different nth powers.

Question. Is it possible that there is no x with the required property?

Answer. The same question was discussed for the square root. Those arguments are still valid, and we have no other (more convincing) ones.

Question. If the degree n is even, then the number $-\sqrt[n]{a}$ also has its nth power equal to a. Why do we prefer the positive x such that $x^n = a$ and reject the negative one?

63 Roots

Answer. This is a generally accepted convention.

Question. If the degree n is odd, then for negative a we can also find an x such that $x^n = a$. For example, $(-2)^3 = -8$. So why do we do not say that the cube root of -8 is -2?

Answer. It is possible to extend our definition to this case (and sometimes people do so), but for simplicity we will consider only non-negative roots of nonnegative numbers. (Otherwise we should consider two cases – odd and even n – all the time.)

Problem 271. Which number is bigger: $\sqrt[10]{2}$ or 1.2?

Problem 272. Compute $\sqrt[7]{0.999}$ to three decimal digits.

Problem 273. Which number is bigger: $\sqrt{2}$ or $\sqrt[3]{3}$?

Problem 274. Which number is bigger: $\sqrt[3]{3}$ or $\sqrt[4]{4}$?

Problem 275. Which number is bigger: $\sqrt{\sqrt{2}}$ or $\sqrt[4]{2}$?

Problem 276. What is $\sqrt[1]{a}$ according to our definition?
Answer. $\sqrt[1]{a} = a$ (for $a \geq 0$).

Now we shall prove some properties of roots.
Problem 277. Prove that (for $a \geq 0$, $b \geq 0$)
$$\sqrt[n]{ab} = \sqrt[n]{a} \cdot \sqrt[n]{b}.$$

Solution. According to the definition of $\sqrt[n]{ab}$ we have to prove that
$$\left(\sqrt[n]{a} \cdot \sqrt[n]{b}\right)^n = ab.$$
Using that
$$(xy)^n = x^n \cdot y^n$$
we get (let $x = \sqrt[n]{a}$, $y = \sqrt[n]{b}$)
$$\left(\sqrt[n]{a} \cdot \sqrt[n]{b}\right)^n = \left(\sqrt[n]{a}\right)^n \cdot \left(\sqrt[n]{b}\right)^n = ab.$$

Problem 278. Prove that (for nonnegative a and b)
$$\sqrt[n]{\frac{a}{b}} = \frac{\sqrt[n]{a}}{\sqrt[n]{b}}.$$

121

Hint. You may use the equation

$$\left(\frac{x}{y}\right)^n = \frac{x^n}{y^n}$$

or the preceding problem.

Problem 279. Prove that for positive a

$$\sqrt[n]{\frac{1}{a}} = \frac{1}{\sqrt[n]{a}}.$$

Problem 280. Prove that for three nonnegative numbers a, b, and c

$$\sqrt[n]{abc} = \sqrt[n]{a} \cdot \sqrt[n]{b} \cdot \sqrt[n]{c}.$$

Solution.

$$\sqrt[n]{abc} = \sqrt[n]{(ab)c} = \sqrt[n]{ab} \cdot \sqrt[n]{c} = \sqrt[n]{a} \cdot \sqrt[n]{b} \cdot \sqrt[n]{c}.$$

The same statement is true for four, five, etc. numbers.

Problem 281. Prove that for nonnegative a

$$\sqrt[n]{a^m} = \left(\sqrt[n]{a}\right)^m.$$

Solution.

$$\sqrt[n]{a^m} = \underbrace{\sqrt[n]{a \cdot a \cdots a}}_{m \text{ times}} = \underbrace{\sqrt[n]{a} \cdot \sqrt[n]{a} \cdots \sqrt[n]{a}}_{m \text{ times}} = \left(\sqrt[n]{a}\right)^m.$$

(We used the statement of the preceding problem.)

Problem 282. There is a flaw in the solution of the preceding problem; find and correct it.

Solution. We assumed that $m \geq 2$; however, the statement makes sense for all integers m (and positive integers n). The cases $m = 0$ and $m = 1$ are trivial. Let us prove it for negative values of m. For example, assume that $m = -3$. Then

$$\sqrt[n]{a^{-3}} = \sqrt[n]{\frac{1}{a^3}} = \frac{1}{\sqrt[n]{a^3}} = \frac{1}{\left(\sqrt[n]{a}\right)^3} = \left(\sqrt[n]{a}\right)^{-3}.$$

Problem 283. Prove that

$$\sqrt[mn]{a} = \sqrt[m]{\sqrt[n]{a}}$$

for any positive integers m, n and for any nonnegative a.

Solution. According to the definition of mnth root we have to prove that

$$\left(\sqrt[m]{\sqrt[n]{a}}\right)^{mn} = a.$$

Indeed,

$$\left(\sqrt[m]{\sqrt[n]{a}}\right)^{mn} = \left(\left(\sqrt[m]{\sqrt[n]{a}}\right)^m\right)^n = \left(\sqrt[n]{a}\right)^n = a.$$

Problem 284. Prove that

$$\sqrt[mn]{a^n} = \sqrt[m]{a}$$

(here m, n are positive integers, $a \geq 0$).

Problem 285. Prove that

$$\sqrt[n]{a^n b} = a\sqrt[n]{b}$$

(n is a positive integer, $a \geq 0$, $b \geq 0$).

64 Non-integer powers

Different properties of roots are hard to remember. The following mnemonic rule may be useful: All of them can be obtained from the known properties of powers if we agree that

$$\sqrt{a} = a^{1/2}, \quad \sqrt[3]{a} = a^{1/3}, \quad \sqrt[4]{a} = a^{1/4} \qquad \text{etc.}$$

For example, the main property of roots (in fact, the definition)

$$\left(\sqrt[n]{a}\right)^n = a$$

now may be rewritten as

$$\left(a^{1/n}\right)^n = a$$

and becomes a special case of the general rule

$$(a^p)^q = a^{pq}$$

where $p = 1/n$, $q = n$.

The property

$$\sqrt[n]{\sqrt[m]{a}} = \sqrt[mn]{a}$$

now may be rewritten as

$$\left(a^{1/m}\right)^{1/n} = a^{1/mn}$$

and can be obtained if we let $p = 1/m$, $q = 1/n$.

Problem 286. Do the same thing for all properties of roots mentioned above (using appropriate properties of powers).

Mnemonic rules are always disappointing, so let us make the status of our rule higher and call it a *definition* of the $1/n$-th power (we may do so because before, we had only integer powers).

Definition. For any integer $n \geq 1$ let

$$a^{1/n} = \sqrt[n]{a}.$$

We immediately observe that this definition does not make us completely happy. For example, we would like to write that

$$a^{\frac{1}{3}} \cdot a^{\frac{1}{3}} = a^{\frac{1}{3}+\frac{1}{3}} = a^{\frac{2}{3}}.$$

(as a special case of the rule $a^m \cdot a^n = a^{m+n}$ where $m = n = \frac{1}{3}$). But we do not know what $a^{2/3}$ is. To fill this gap we *define* $a^{2/3}$ as $(a^{1/3})^2$ and, in general, $a^{m/n}$ as $(a^{1/n})^m$ or, in other words, as $(\sqrt[n]{a})^m$. So we come to the following

Definition. For any integer m and for any positive integer n the expression $a^{m/n}$ is defined as follows:

$$a^{\frac{m}{n}} = \left(\sqrt[n]{a}\right)^m.$$

The careful reader would mention that there is some cheating in this definition. Indeed,

$$a^{\frac{10}{15}} \text{ is defined as } \left(\sqrt[15]{a}\right)^{10}$$

and at the same time

$$a^{\frac{2}{3}} \text{ is defined as } \left(\sqrt[3]{a}\right)^2$$

At the same time $\frac{10}{15} = \frac{2}{3}$ so $a^{10/15}$ must be equal to $a^{2/3}$. So the correctness of our definition requires that

$$\left(\sqrt[15]{a}\right)^{10} = \left(\sqrt[3]{a}\right)^{2}.$$

Problem 287. Prove this fact.

Solution.

$$\left(\sqrt[3\cdot5]{a}\right)^{2\cdot5} = \left(\left(\sqrt[5]{\sqrt[3]{a}}\right)^{5}\right)^{2} = \left(\sqrt[3]{a}\right)^{2}.$$

Problem 288. Prove that reducing common factors in the fraction $\frac{m}{n}$ does not change the value of the expression $a^{\frac{m}{n}}$ (see the definition above).

Hint. In the preceding problem the common factor 5 was reduced in the fraction $\frac{10}{15}$.

Now the properties of powers that we know for integer powers should be checked for arbitary rational powers (where the exponent is a ratio of any integers).

Problem 289. Prove that

$$a^{p} \cdot a^{q} = a^{p+q}$$

for any rational p and q.

Solution. For example, let $p = 2/5$, $q = 3/7$. We have to check that

$$a^{\frac{2}{5}} \cdot a^{\frac{3}{7}} = a^{\frac{2}{5}+\frac{3}{7}}.$$

Let us find a common denominator for $2/5$ and $3/7$:

$$\frac{2}{5} = \frac{14}{35}, \quad \frac{3}{7} = \frac{15}{35}.$$

As we know already,

$$a^{\frac{2}{5}} = a^{\frac{14}{35}}, \quad a^{\frac{3}{7}} = a^{\frac{15}{35}},$$

therefore

$$a^{\frac{2}{5}} \cdot a^{\frac{3}{7}} = a^{\frac{14}{35}} \cdot a^{\frac{15}{35}} = \left(\sqrt[35]{a}\right)^{14} \cdot \left(\sqrt[35]{a}\right)^{15} =$$
$$= \left(\sqrt[35]{a}\right)^{14+15} = a^{\frac{14+15}{35}} = a^{\frac{14}{35}+\frac{15}{35}} = a^{\frac{2}{5}+\frac{3}{7}}.$$

Problem 290. Prove that

$$(ab)^{m/n} = a^{m/n} \cdot b^{m/n}.$$

Problem 291. Prove that

$$(a^p)^q = a^{pq}$$

for any rational p and q.

Solution. Let us start with the case of integer q and arbitrary rational $p = m/n$. In this case

$$(a^p)^q = \left(a^{\frac{m}{n}}\right)^q = \left(\left(\sqrt[n]{a}\right)^m\right)^q = \left(\sqrt[n]{a}\right)^{mq} = a^{\frac{mq}{n}} = a^{pq}.$$

Assume now that $q = 1/k$ for some integer k and that $p = m/n$. Then

$$(a^p)^q = \left(a^{\frac{m}{n}}\right)^{\frac{1}{k}} = \sqrt[k]{a^{\frac{m}{n}}} = \sqrt[k]{\left(\sqrt[n]{a}\right)^m}$$

Let us denote $\sqrt[n]{a}$ as b and continue this chain of equalities:

$$\cdots = \sqrt[k]{b^m} = \left(\sqrt[k]{b}\right)^m = \left(\sqrt[k]{\sqrt[n]{a}}\right)^m = \left(\sqrt[kn]{a}\right)^m = a^{\frac{m}{kn}} = a^{\frac{m}{n} \cdot \frac{1}{k}} = a^{pq}.$$

Finally, for an arbitrary $q = l/k$ we have

$$(a^p)^q = (a^p)^{\frac{l}{k}} = \left(\sqrt[k]{a^p}\right)^l = \left(\left(a^p\right)^{\frac{1}{k}}\right)^l = \left(a^{\frac{p}{k}}\right)^l = a^{\frac{pl}{k}} = a^{pq}.$$

We used that

$$(a^p)^{\frac{1}{k}} = a^{p \cdot \frac{1}{k}}$$

and then we used that

$$\left(a^{\frac{p}{k}}\right)^l = a^{\frac{p}{k} \cdot l}.$$

These two special cases of the statement of the problem are considered already.

Problem 292. Prove that for $a > 1$ the value of a^p increases when p increases. Prove that for $0 < a < 1$ the value of a^p *decreases* when p increases.

Hint. When comparing two values of p, find the common denominator. Do not forget that p may be negative (and the statement of the problem remains true).

This problem shows a possible way to extend the definition of a^x to the irrational values of x. For example, we may try to define

$$2^{\sqrt{2}}$$

as a number that is bigger than any of the numbers $2^{p/q}$ when $p/q < \sqrt{2}$ but smaller than any of the numbers $2^{p/q}$ when $p/q > \sqrt{2}$. Of course, to make this definition correct we must prove that such a number exists and is unique, but these topics belong to the scope of calculus.

Problem 293. How do you think one should define $\sqrt[1/2]{a}$ or $\sqrt[-1/2]{a}$?

Answer. As a^2 and a^{-2}.

65 Proving inequalities

Almost all the inequalities in this section in principle could be proved by "brute force" if we computed the values of all the expressions. But we shall look for a better way.

Problem 294. Prove that

$$\frac{1}{2} < \frac{1}{101} + \frac{1}{102} + \cdots + \frac{1}{200} < 1.$$

Solution. Each of 100 terms of the sum is between $\frac{1}{200}$ and $\frac{1}{100}$. If all terms were equal to $\frac{1}{200}$, the sum would be equal to $\frac{1}{2}$; if all terms were equal to $\frac{1}{100}$, the sum would be equal to 1.

Problem 295. Prove that

$$\frac{1}{2} < 1 - \frac{1}{2} + \frac{1}{3} - \frac{1}{4} + \cdots + \frac{1}{199} - \frac{1}{200} < 1.$$

Solution. The left inequality can be proved by grouping the terms with parentheses as

$$\left(1 - \frac{1}{2}\right) + \left(\frac{1}{3} - \frac{1}{4}\right) + \cdots + \left(\frac{1}{199} - \frac{1}{200}\right).$$

Here the first parenthesized grouping is equal to $1/2$, and all the others are positive.

To get the right inequality we rewrite the expression as

$$1 - \left(\frac{1}{2} - \frac{1}{3}\right) - \left(\frac{1}{4} - \frac{1}{5}\right) - \cdots - \left(\frac{1}{198} - \frac{1}{199}\right) - \frac{1}{200}.$$

Here all parenthesized groupings are positive.

Remark. In fact the preceding two problems coincide in a sense:

$$\frac{1}{101} + \frac{1}{102} + \cdots + \frac{1}{200} = 1 - \frac{1}{2} + \frac{1}{3} - \frac{1}{4} + \cdots + \frac{1}{199} - \frac{1}{200}.$$

Problem 296. Prove this coincidence.

Solution. Indeed,

$$\frac{1}{101} + \cdots + \frac{1}{200} =$$

$$= \left(1 + \frac{1}{2} + \frac{1}{3} + \cdots + \frac{1}{200}\right) - \left(1 + \frac{1}{2} + \frac{1}{3} + \cdots + \frac{1}{100}\right) =$$

$$= \left(1 + \frac{1}{2} + \frac{1}{3} + \cdots + \frac{1}{200}\right) - 2 \cdot \left(\frac{1}{2} + \frac{1}{4} + \frac{1}{6} + \frac{1}{8} + \cdots + \frac{1}{200}\right) =$$

$$= 1 - \frac{1}{2} + \frac{1}{3} - \frac{1}{4} + \cdots + \frac{1}{199} - \frac{1}{200}.$$

Problem 297. Prove that $(1.01)^{100} \geq 2$.

Solution. By definition

$$(1.01)^{100} = \underbrace{(1 + 0.01)(1 + 0.01) \cdots (1 + 0.01)}_{100 \text{ factors}}$$

What happens if we remove the parentheses? We get a sum of many products (each term of this sum is a product of 100 numbers – one for each parenthesized expression). One of the terms is 1 (a product of all the ones). Among other terms there are terms being products of 99 ones and only one 0.01. We have 100 terms of this type (because the 0.01 term could be taken from any of the parenthesized expressions). The value of such a term is 0.01. There are other terms also (equal to 0.01^2, 0.01^3, etc.) but even if we omit them we get the sum

$$1 + 100 \cdot 0.01 = 2.$$

Another solution to the same problem goes as follows:

$$
\begin{aligned}
1.01^2 &= 1.0201 > 1.02 \\
1.01^3 &= 1.01^2 \cdot 1.01 > 1.02 \cdot 1.01 = (1+0.02)(1+0.01) = \\
&= 1 + 0.02 + 0.01 + 0.02 \cdot 0.01 > 1.03 \\
1.01^4 &= 1.01^3 \cdot 1.01 > 1.03 \cdot 1.01 = (1+0.03)(1+0.01) = \\
&= 1 + 0.03 + 0.01 + 0.03 \cdot 0.01 > 1.04 \\
1.01^5 &> 1.05 \\
1.01^6 &> 1.06 \\
&\cdots \\
1.01^{99} &> 1.99 \\
1.01^{100} &> 2 \,.
\end{aligned}
$$

Problem 298. Prove that

$$
1 + \frac{1}{4} + \frac{1}{9} + \frac{1}{16} + \cdots + \frac{1}{100^2} < 2 \,.
$$

Solution.

$$
\begin{aligned}
\frac{1}{4} = \frac{1}{2^2} &< \frac{1}{1 \cdot 2} = \frac{1}{1} - \frac{1}{2} \\
\frac{1}{9} = \frac{1}{3^2} &< \frac{1}{2 \cdot 3} = \frac{1}{2} - \frac{1}{3} \\
\frac{1}{16} = \frac{1}{4^2} &< \frac{1}{3 \cdot 4} = \frac{1}{3} - \frac{1}{4} \\
&\cdots \\
\frac{1}{100^2} &< \frac{1}{99 \cdot 100} = \frac{1}{99} - \frac{1}{100} \,,
\end{aligned}
$$

hence (adding all the inequalities),

$$
\begin{aligned}
1 + \frac{1}{4} + \frac{1}{9} + \frac{1}{16} + \cdots + \frac{1}{100^2} &< \\
< 1 + \left(1 - \frac{1}{2}\right) + \left(\frac{1}{2} - \frac{1}{3}\right) + \left(\frac{1}{3} - \frac{1}{4}\right) + \cdots + \left(\frac{1}{99} - \frac{1}{100}\right) &= \\
= 1 + 1 - \frac{1}{100} < 2 \,.
\end{aligned}
$$

Here the terms $-\dfrac{1}{2}$ and $\dfrac{1}{2}$, $-\dfrac{1}{3}$ and $\dfrac{1}{3}$, etc. cancel out.

Problem 299. Which is bigger: 1000^{2000} or 2000^{1000}?

Problem 300. Prove that

$$1 + \frac{1}{2} + \frac{1}{3} + \cdots + \frac{1}{1{,}000{,}000} < 20.$$

Prove that

$$1 + \frac{1}{2} + \frac{1}{3} + \cdots + \frac{1}{n} > 20.$$

for some n.

Hint. In the expression

$$\left(\frac{1}{2}\right) + \left(\frac{1}{3} + \frac{1}{4}\right) + \left(\frac{1}{5} + \frac{1}{6} + \frac{1}{7} + \frac{1}{8}\right) + \left(\frac{1}{9} + \frac{1}{10} + \cdots + \frac{1}{16}\right) + \cdots$$

each expression in parentheses is between $1/2$ and 1 (compare with the first problem of this section).

66 Arithmetic and geometric means

The *arithmetic mean* (pronounced "arithmEtic", not "arIthmetic") of two numbers a and b is defined as $\dfrac{a+b}{2}$, that is, as half of their sum. The corresponding point on the real line is the midpoint of the segment with endpoints a and b.

Problem 301. Prove this fact.

Solution. Without loss of generality we may assume that $a < b$. In this case point a is on the left of point b.

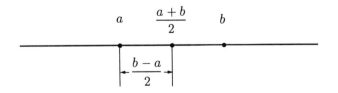

The distance between these points is equal to $b - a$; if we add to a one-half of this distance we get

$$a + \frac{b - a}{2} = \frac{2a + b - a}{2} = \frac{a + b}{2}.$$

Problem 302. The arithmetic mean of two numbers 1 and a is equal to 7. Find a.

The *geometric mean* of two nonnegative numbers a and b is defined as the square root of their product, $\sqrt{ab}$. We restrict ourselves to nonnegative a and b; if a and b have different signs, their product is negative and the square root is undefined. If both numbers are negative, then $\sqrt{ab}$ is defined, but it would be strange to call the positive number $\sqrt{ab}$ a geometric mean of two negative numbers!

Problem 303. The geometric mean of two numbers 1 and a is equal to 7. Find a.

Problem 304. (a) Find the side of a square having the same perimeter as a rectangle with sides a and b. (b) Find the side of a square having the same area as a rectangle with sides a and b.

Problem 305. We have already heard about arithmetic and geometric progressions, and now we learn the terms "arithmetic mean" and "geometric mean". Can you explain this coincidence of terms?

Solution. The sequence

$$a, \quad \text{(the arithmetic mean of } a \text{ and } b), \quad b$$

is an arithmetic progression while the sequence

$$a, \quad \text{(the geometric mean of } a \text{ and } b), \quad b$$

is a geometric progression.

One more way to define the arithmetic and geometric mean:

- The arithmetic mean of a and b is a number x such that

$$x - a = b - x;$$

- The geometric mean of a and b is a number x such that

$$\frac{x}{a} = \frac{b}{x} \quad \text{(for } a, b > 0\text{)}.$$

67 The geometric mean does not exceed the arithmetic mean

Problem 306. Prove that for nonnegative a and b

$$\sqrt{ab} \leq \frac{a+b}{2}$$

Solution. To compare nonnegative numbers $\sqrt{ab}$ and $\frac{a+b}{2}$ let us compare their squares and prove that

$$ab \leq \left(\frac{a+b}{2}\right)^2.$$

Taking into account that

$$\left(\frac{a+b}{2}\right)^2 = \frac{(a+b)^2}{4}$$

we have to prove that

$$ab \leq \frac{(a+b)^2}{4}$$

or, in other words, that $4ab \leq (a+b)^2$, or $4ab \leq a^2 + 2ab + b^2$, or $0 \leq a^2 - 2ab + b^2$.

It is easy to recognize $(a-b)^2$ as the right-hand side of this inequality, therefore it is proved (a square is always nonnegative).

Problem 307. When is the arithmetic mean of two numbers equal to their geometric mean?

Solution. As we see from the solution of the preceeding problem, this happens if and only if $(a-b)^2 = 0$, that is, if $a = b$.

68 Problems about maximum and minimum

Problem 308. (a) What is the maximum value of the product of two nonnegative numbers whose sum is a fixed positive number c ? (b) What is its minimum value?

Solution. (a) The arithmetic mean of these numbers is $c/2$, so their geometric mean cannot exceed $c/2$, and its square (that is, the product of the numbers) never exceeds $c^2/4$. This maximum value is achieved when the numbers are equal.

132

(b) The minimum value is zero (one of the numbers is zero, the other one is equal to c).

Problem 309. What are the maximum and minimum values of the sum of two nonnegative numbers whose product is a fixed $c > 0$?

Solution. The geometric mean of these numbers is $\sqrt{c}$. Therefore their arithmetic mean is not less than $\sqrt{c}$ and their sum (which is two times bigger) is not less than $2\sqrt{c}$. This value is achieved if the numbers are equal. The maximum value does not exist (the sum may be arbitrarily large if one of the numbers is close to zero and the other one is very large).

Remark. As you may remember, we have met the two last problems earlier when speaking about maximum and minimum values of quadratic polynomials.

Problem 310. What is the maximum possible area of a rectangular piece of land if you may enclose it with only 120 m of fence?

Problem 311. What is the maximum possible area of a rectangular piece of land near the (straight) sea shore if you may enclose it with only 120 m of fence? (You don't need the fence on the shore or in the water.)

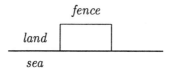

Solution. Imagine the symmetric fence in the sea:

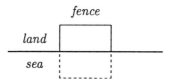

We get a rectangle (half in the water) with perimeter equal to 240 m. Its area will be maximal if it is a square with side 60 m. In this case the area is equal to $3600\,\mathrm{m}^2$. The real area (on the shore) is half of this and equals $1800\,\mathrm{m}^2$; the real fence contains of segments of length 30, 60, and 30 meters.

Problem 312. What is the maximum value of the product ab if a and b are nonnegative numbers such that $a + 2b = 3$?

Solution. It is easier to say when the product of two nonnegative numbers a and $2b$ (whose sum equals 3) is maximal. It is maximal when these numbers are equal, that is, $a = 2b = 3/2$. The product of a and b is half the product of a and $2b$; its maximum value is

$$\frac{3}{2} \cdot \frac{3}{4} = \frac{9}{8}.$$

69 Geometric illustrations

The inequality

$$\sqrt{ab} \le \frac{a+b}{2}$$

can be rewritten as

$$2\sqrt{ab} \le a + b$$

and then, after squaring, as

$$4ab \le (a + b)^2$$

The last inequality can be illustrated as follows: Four rectangles $a \times b$ can be put into the square with side $a + b$ (and some space in the middle of the square remains, if $a \ne b$).

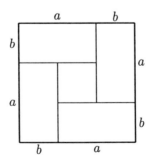

Problem 313. How much free space remains? Compare the result

with the algebraic proof of the inequality given above.

Another illustration is as follows. Consider the bisector of a right angle, and two triangles with sides a and b parallel to the sides of the angle:

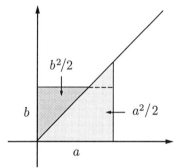

Their areas are $a^2/2$ and $b^2/2$. Together these triangles cover a rectangle with sides a and b; therefore

$$ab \leq \frac{a^2 + b^2}{2}.$$

To see that this illustrates the inequality between the arithmetic and the geometric mean, substitute $\sqrt{c}$ and $\sqrt{d}$ for a and b; you get

$$\sqrt{c} \cdot \sqrt{d} \leq \frac{c + d}{2}.$$

Remark. You may use almost any curve instead of the bisector – and obtain many other inequalities, if you are able to compute the areas of triangles formed by curves.

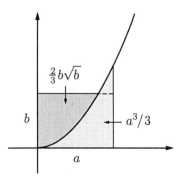

For example, for the curve $y = x^2$ you get (as calculus experts say) two "triangles" having areas $a^3/3$ and $\frac{2}{3}b\sqrt{b}$. So the inequality obtained is

$$ab \le \frac{a^3}{3} + \frac{2}{3}b\sqrt{b}.$$

This is true for any nonnegative a and b.

70 The arithmetic and geometric means of several numbers

The arithmetic mean of three numbers is defined as $\dfrac{a+b+c}{3}$; the geometric mean is defined as $\sqrt[3]{abc}$ (we assume that $a, b, c \ge 0$). Similar definitions are given for four, five, etc. numbers; the arithmetic mean of $a_1, \ldots, a_n$ is

$$\frac{a_1 + \cdots + a_n}{n};$$

the geometric mean is

$$\sqrt[n]{a_1 \cdot a_2 \cdots a_n}.$$

The inequality between the arithmetic and the geometric means can be generalized for the case of n numbers:

$$\sqrt[n]{a_1 \cdots a_n} \le \frac{a_1 + \cdots + a_n}{n}.$$

As for the case of two numbers (see above), equality is possible only if all numbers are equal.

Before proving this inequality we shall derive some of its consequences.

Problem 314. Using this inequality, prove that if $a_1, \ldots, a_n$ are nonnegative numbers and $a_1 + a_2 + \cdots + a_n \le n$, then $a_1 \cdot a_2 \cdots a_n \le 1$.

Solution.

$$a_1 + \cdots + a_n \le n \quad \Longrightarrow \quad \frac{a_1 + \cdots + a_n}{n} \le 1 \quad \Longrightarrow$$

$$\Longrightarrow \quad \sqrt[n]{a_1 \cdot a_2 \cdots a_n} \le \frac{a_1 + \cdots + a_n}{n} \le 1 \quad \Longrightarrow \quad a_1 \cdot a_2 \cdots a_n \le 1.$$

In the following two problems you may also use the inequality between the arithmetic and the geometric means without proof.

Problem 315. Prove that the product of n nonnegative numbers with a fixed sum is at a maximum when all the numbers are equal.

Problem 316. Prove that the sum of n nonnegative numbers with a given product is at a minimum when all the numbers are equal.

There are different proofs of the inequality between the arithmetic mean and the geometric mean of n numbers. Unfortunately, the most natural of them uses calculus (the notion of a derivative or something else). We shall avoid that, but our proofs will be tricky.

Problem 317. Prove the inequality between arithmetic and geometric means for $n = 4$.

Solution. We have four nonnegative numbers. During the proof we will change them but keep their sum unchanged. (Therefore, their arithmetic mean will be unchanged.) Their product will change and we'll keep track of how.

1. Replace a and b by two numbers equal to $\dfrac{a+b}{2}$; so we make a transition

$$a, \ b, \ c, \ d \ \longrightarrow \ \frac{a+b}{2}, \ \frac{a+b}{2}, \ c, \ d \, .$$

The sum remains unchanged while the product increases (when $a \neq b$) or remains the same (if $a = b$); two factors c and d do not change and the product of two numbers with fixed sum $a+b$ is maximal when numbers are equal (see above).

2. Do the same with c and d:

$$\frac{a+b}{2}, \ \frac{a+b}{2}, \ c, \ d \ \longrightarrow \ \frac{a+b}{2}, \ \frac{a+b}{2}, \ \frac{c+d}{2}, \ \frac{c+d}{2} \, .$$

The sum remains unchanged, the product increases or remains the same (if $c = d$).

3. We have balanced the first and the second pair; now we balance numbers of different pairs:

$$\frac{a+b}{2}, \ \frac{a+b}{2}, \ \frac{c+d}{2}, \ \frac{c+d}{2} \ \longrightarrow$$
$$\longrightarrow \ \frac{a+b+c+d}{4}, \ \frac{a+b}{2}, \ \frac{a+b+c+d}{4}, \ \frac{c+d}{2} \, .$$

and, finally,

$$\frac{a+b+c+d}{4}, \frac{a+b}{2}, \frac{a+b+c+d}{4}, \frac{c+d}{2} \longrightarrow$$
$$\longrightarrow \frac{a+b+c+d}{4}, \frac{a+b+c+d}{4}, \frac{a+b+c+d}{4}, \frac{a+b+c+d}{4}.$$

So ultimately we replaced numbers

$$a, \ b, \ c, \ d$$

by numbers

$$S, \ S, \ S, \ S$$

where

$$S = \frac{a+b+c+d}{4}$$

is the arithmetic mean, and their product increased (or at least did not decrease), so

$$a \cdot b \cdot c \cdot d \leq S \cdot S \cdot S \cdot S$$

or

$$\sqrt[4]{abcd} \leq S.$$

The inequality is proved!

Problem 318. Prove that the inequality between the arithmetic and the geometric means of four numbers (see the preceding problem) becomes an equality only if all numbers are equal.

Hint. Look at the solution of the preceding problem; the final equality is possible only if at all stages numbers being balanced are equal.

Problem 319. Prove the inequality between arithmetic and geometric means for $n = 8$.

Solution. Do the same trick as in the preceding proof: balance numbers in four pairs, then (between pairs) in two quadruples, and then all eight.

Problem 320. Prove the inequality between arithmetic and geometric means for $n = 3$.

Solution. We reduce this problem to the case $n = 4$ by the following method: besides three given numbers a, b, c consider the fourth number, namely, their geometric mean. So we get four numbers

$$a, b, c, \sqrt[3]{abc}$$

and then use the inequality for $n = 4$; we get

$$\sqrt[4]{abc\sqrt[3]{abc}} \leq \frac{a + b + c + \sqrt[3]{abc}}{4} .$$

The left-hand side expression turns out to be equal to $\sqrt[3]{abc}$. To verify this, compute the fourth powers of both (nonnegative) numbers; we get

$$\left(\sqrt[4]{abc\sqrt[3]{abc}} \right)^4 = abc \sqrt[3]{abc}$$

and

$$\left(\sqrt[3]{abc} \right)^4 = \left(\sqrt[3]{abc} \right)^3 \sqrt[3]{abc} = abc \sqrt[3]{abc},$$

which is the same. So we can rewrite the inequality we have as

$$\sqrt[3]{abc} \leq \frac{a + b + c + \sqrt[3]{abc}}{4}$$

and then

$$4\sqrt[3]{abc} \leq a + b + c + \sqrt[3]{abc},$$
$$3\sqrt[3]{abc} \leq a + b + c,$$
$$\sqrt[3]{abc} \leq \frac{a + b + c}{3} .$$

That is what we want.

Problem 321. Using the inequality between arithmetic and geometric means for $n = 8$, prove it for $n = 7$.

Problem 322. Prove the inequality between arithmetic and geometric means for $n = 6$.

Hint. Recall the solution of the preceding problem.

Problem 323. Prove the inequality between arithmetic and geometric means for all integer $n \geq 2$.

Hint. Prove it for $n = 2, 4, 8, 16, 32, \ldots$ and then all integers in between them.

Problem 324. Prove that the inequality between arithmetic and geometric means becomes an equality only if all numbers are equal.

Another proof of the inequality between arithmetic and geometric means goes as follows. First of all we mention that if all numbers $a_1, \ldots, a_n$ are multiplied by the same constant (for example, if all numbers become three times bigger) then both the arithmetic and geometric means are multiplied by the same constant and the relation between them remains unchanged. Therefore, proving the inequality between them, we may multiply all numbers by some constant and assume without loss of generality that their arithmetic mean is equal to 1. Thus, it is enough to prove

$$a_1, \ldots, a_n \geq 0, \; a_1 + \cdots + a_n = n \; \Rightarrow \; a_1 \cdots a_n \leq 1.$$

Let's try.

A. For the case of two numbers: If the sum of two numbers is equal to 2, then these numbers can be represented as $1 + h$ and $1 - h$ and their product is $(1 + h)(1 - h) = 1 - h^2 \leq 1$.

B. Let us consider now the case of three numbers. Assume that the sum of three nonnegative numbers a, b, c is 3. If not all of a, b, c are equal to 1 (the latter case is trivial) some of them must be greater than 1 and some must be smaller. Assume, for instance, that $a < 1$ and $b > 1$. Then $a - 1 < 0$, $b - 1 > 0$ and the product

$$(a - 1)(b - 1) = ab - a - b + 1$$

is negative, so

$$ab + 1 < a + b.$$

Because

$$(a + b) + c = 3$$

we have

$$ab + 1 + c < (a + b) + c = 3$$

and

$$ab + c < 2.$$

140

Look – we now have two numbers ab and c, their sum is less than 2 and we have to prove that their product does not exceed 1. For two numbers we already know this fact from part **A**.

The careful reader may ask why we refer to part **A** where we proved that if the sum of two numbers is *equal* to 2 then the product does not exceed 1, and now the sum of two numbers is *smaller* than 2, not *equal* to 2. But this is not a big problem; if the sum is smaller than 2 we may increase one of the numbers and make the sum equal to 2; if the increased product does not exceed 1 then the original product also does not exceed 1.

C. Now assume that $n = 4$; we have to prove that

$$a,\ b,\ c,\ d \geq 0,\ a + b + c + d = 4 \implies abcd \leq 1.$$

Again we may assume without loss of generality that one of the numbers, say a, is less than 1 and the other, say b, is greater than 1. Then

$$ab + 1 < a + b, \quad (a + b) + c + d = 4,$$

therefore

$$ab + 1 + c + d < 4, \quad ab + c + d < 3.$$

And again it remains to prove that if a sum of three nonnegative numbers ab, c, and d is less than 3 then their product does not exceed 1 – and this is already proved.

The same argument can be applied for $n = 5, 6$, etc.

The next, third proof of the inequality between the arithmetic and the geometric means of three numbers is probably the shortest – but it looks mysterious.

We start from the identity

$$a^3 + b^3 + c^3 - 3abc = \frac{1}{2}(a + b + c)((a - b)^2 + (b - c)^2 + (a - c)^2)$$

which can be checked by a direct computation (perform the operations on the right-hand side). You see that if a, b, c are nonnegative then the right-hand side (and therefore left-hand side) is nonnegative, that is,

$$abc \leq \frac{a^3 + b^3 + c^3}{3}.$$

It remains to substitute $\sqrt[3]{p}$, $\sqrt[3]{q}$, $\sqrt[3]{r}$, for a, b, and c and you get

$$\sqrt[3]{pqr} \leq \frac{p+q+r}{3}$$

which concludes the third proof.

Here is one more proof of the inequality between the arithmetic and geometric means. Let us prove that the product of n nonnegative numbers is the maximum when all the numbers are equal. As we have seen, it is easy to prove this fact for $n = 2$. Assume that for some n the product of n equal numbers with a given sum S is *not* the maximum, and that some other numbers $a_1, a_2, \ldots, a_n$ – not all of them equal – provide this maximum. Assume, for example, that $a_1 \neq a_2$. Then replacing both of a_1 and a_2 by their arithmetic mean, we do not change the sum, but the product increases. So we get a contradiction with our assumption that the product was the maximum.

Problem 325. There is a gap in this argument – find it.

Solution. We assumed that numbers $a_1, a_2, \ldots, a_n$ providing the maximum value of the product (for nonnegative numbers with fixed sum) do exist. This fact needs to be proved. In fact it can be proved using calculus methods, but this goes beyond the scope of the book.

Problem 326. Assume that $a_1, \ldots, a_n$ are positive numbers. Prove that

$$\frac{a_1}{a_2} + \frac{a_2}{a_3} + \cdots + \frac{a_{n-1}}{a_n} + \frac{a_n}{a_1} \geq n.$$

Problem 327. Prove that

$$\sqrt[3]{ab^2} \leq \frac{a+2b}{3}.$$

Problem 328. Find the minimal value of $a + b$ if a and b are nonnegative numbers and $ab^2 = 1$.

Problem 329. Prove the inequality

$$\sqrt[6]{a} \cdot \sqrt[3]{b} \cdot \sqrt{c} \leq \frac{a + 2b + 3c}{6}$$

for any nonnegative a, b, and c.

Problem 330. Prove the inequality

$$\sqrt[3]{abc} \le \frac{a + 2b + 3c}{3\sqrt[3]{6}}$$

Problem 331. Prove that

$$\left(1 + \frac{1}{10}\right)^{10} < \left(1 + \frac{1}{11}\right)^{11}.$$

Solution. The left-hand side $\left(1 + \frac{1}{10}\right)^{10}$ is a product of 10 factors each equal to $\left(1 + \frac{1}{10}\right)$. We may consider it also as a product of 11 factors, one of them equal to 1 and ten others equal to $\left(1 + \frac{1}{10}\right)$. Comparing this product with the product in the right-hand side where we also have 11 factors but all of them are equal to $\left(1 + \frac{1}{11}\right)$, we see that the sum of all factors are the same in both cases (namely 12). But in the right-hand side all factors are equal, so the product is bigger.

Problem 332. Prove that

$$\left(1 + \frac{1}{10}\right)^{11} > \left(1 + \frac{1}{11}\right)^{12}.$$

Hint. The right-hand side may be considered as a product of 11 factors – one equal to

$$\left(1 + \frac{1}{11}\right)^{2} = \left(1 + \frac{2}{11} + \frac{1}{11^2}\right)$$

and the others equal to $\left(1 + \frac{1}{11}\right)$. The left-hand side is also a product of 11 factors (but the factors are equal). It is enough to show that the sum of all factors in the left-hand side is bigger than the sum of all factors in the right-hand side and then use the inequality between arithmetic and geometric means.

Problem 333. Write down the four numbers mentioned in the two preceding problems in ascending order.

71 The quadratic mean

The *quadratic mean* of two nonnegative numbers a and b is defined as a nonnegative number whose square is the arithmetic mean of a^2 and b^2, that is, as

$$\sqrt{\frac{a^2 + b^2}{2}}.$$

Problem 334. This definition uses the arithmetic mean. What happens if the arithmetic mean is replaced by the geometric mean?

Problem 335. Prove that the quadratic mean of two nonnegative numbers a and b is not less than their arithmetic mean:

$$\sqrt{\frac{a^2 + b^2}{2}} \geq \frac{a + b}{2}.$$

(For example, the quadratic mean of 0 and a is $a/\sqrt{2}$ and their arithmetic mean is $a/2$.

Solution. Comparing the squares, we need to prove that

$$\frac{a^2 + b^2}{2} \geq \frac{(a + b)^2}{4}.$$

Multiplying by 4 and using the square-of-the-sum formula, we get

$$2(a^2 + b^2) \geq a^2 + b^2 + 2ab$$

or

$$a^2 + b^2 \geq 2ab, \quad a^2 + b^2 - 2ab \geq 0.$$

Here the left-hand side is a square of $(a - b)$ and, therefore, is always nonnegative.

Problem 336. For which a and b is the arithmetic mean equal to the quadratic mean?

Problem 337. Prove that the geometric mean does not exceed the quadratic mean.

The geometric illustration of the inequality between the arithmetic mean and the quadratic mean can be given as follows.

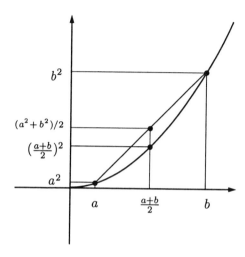

Draw the graph $y = x^2$ and consider two points $\langle a, a^2 \rangle$ and $\langle b, b^2 \rangle$ on this graph. Connect these points with a segment. The middle point of this segment has coordinates that are the arithmetic means of the coordinates of the endpoints, that is,

$$\left(\frac{a+b}{2}, \quad \frac{a^2+b^2}{2} \right).$$

Look at the picture; you see that this point is higher than the graph point with the same x-coordinate

$$\left(\frac{a+b}{2}, \quad \left(\frac{a+b}{2} \right)^2 \right)$$

so the y-coordinate of the first point is bigger than the y-coordinate of the second one:

$$\left(\frac{a+b}{2} \right)^2 \leq \frac{a^2+b^2}{2},$$

$$\frac{a+b}{2} \leq \sqrt{\frac{a^2+b^2}{2}}.$$

This argument may be considered as a proof of the inequality between arithmetic and quadratic means if we believe that the graph of $y = x^2$

is "concave" (that is, the curve goes under the chord connecting any two points).

Problem 338. Turning the graph $y = x^2$ around (that is, exchanging x- and y-axes), we get the graph of $y = \sqrt{x}$, which goes *above* any of its chords. What inequality corresponds to this fact?

Answer.

$$\sqrt{\frac{a+b}{2}} \geq \frac{\sqrt{a} + \sqrt{b}}{2}.$$

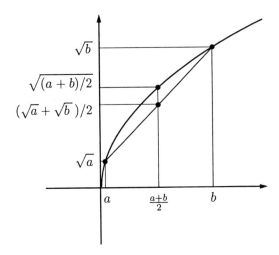

Now we know that for any nonnegative a and b

$$\sqrt{ab} \leq \frac{a+b}{2} \leq \sqrt{\frac{a^2 + b^2}{2}}.$$

For any of these three expressions, let us draw in the coordinate plane the set of all points $\langle a, b \rangle$ where this type of mean value does not exceed 1:

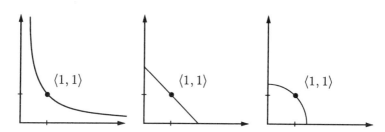

If we put all of them in one picture, we see that the bigger expression corresponds to the smaller set (as it should).

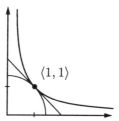

Problem 339. Prove the inequality between the arithmetic mean and the quadratic mean for three numbers:

$$\frac{a+b+c}{3} \leq \sqrt{\frac{a^2+b^2+c^2}{3}}.$$

Problem 340. (a) The sum of two nonnegative numbers is 2. What is the minimum value of the sum of their squares?

(b) The same question for three numbers.

72 The harmonic mean

The *harmonic mean* of two positive numbers a and b is defined (see above) as the number whose inverse is the arithmetic mean of the inverses of a and b, that is, as

$$\frac{1}{\left(\dfrac{1}{a}+\dfrac{1}{b}\right)/2}.$$

Problem 341. Prove that the harmonic mean does not exceed the geometric mean.

Solution. The inverse of the harmonic mean is the arithmetic mean of $1/a$ and $1/b$; the inverse of the geometric mean is the geometric mean of $1/a$ and $1/b$; so it is enough to recall the inequality between the arithmetic and geometric means (the inverse of the bigger number is smaller).

Problem 342. The numbers $a_1, \ldots, a_n$ are positive. Prove that

$$(a_1 + \cdots + a_n)\left(\frac{1}{a_1} + \cdots + \frac{1}{a_n}\right) \geq n^2.$$

Solution. The desired inequality may be rewritten as

$$\frac{a_1 + \cdots + a_n}{n} \geq \frac{1}{\left(\dfrac{1}{a_1} + \cdots + \dfrac{1}{a_n}\right)/n}$$

that is, we have to prove that the arithmetic mean of n numbers is greater than or equal to its harmonic mean. This becomes clear if we put the geometric mean between them:

$$\frac{a_1 + \cdots + a_n}{n} \geq \sqrt[n]{a_1 \cdots a_n} =$$

$$= \frac{1}{\sqrt[n]{\dfrac{1}{a_1} \cdots \dfrac{1}{a_n}}} \geq \frac{1}{\left(\dfrac{1}{a_1} + \cdots + \dfrac{1}{a_n}\right)/n};$$

the last inequality follows from the inequality between the arithmetic and geometric mean of the numbers $1/a_1, \ldots, 1/a_n$.

Another solution uses the following trick. Our inequality becomes a consequence of the so-called Cauchy–Schwarz inequality

$$(p_1 q_1 + \cdots + p_n q_n)^2 \leq (p_1^2 + \cdots + p_n^2) \cdot (q_1^2 + \cdots + q_n^2)$$

if we substitute $\sqrt{a_i}$ for p_i and $\dfrac{1}{\sqrt{a_i}}$ for q_i.

Therefore, it remains to prove the Cauchy–Schwarz inequality. Consider the following quadratic polynomial (where x is considered to be a variable and p_i and q_i are constants):

$$(p_1 + q_1 x)^2 + (p_2 + q_2 x)^2 + \cdots + (p_n + q_n x)^2.$$

If we remove the parentheses and collect terms with x^2, with x, and without x, we get the polynomial

$$Ax^2 + Bx + C$$

where

$$
\begin{aligned}
A &= q_1^2 + q_2^2 + \cdots + q_n^2, \\
B &= 2(p_1 q_1 + p_2 q_2 + \cdots + p_n q_n), \\
C &= p_1^2 + p_2^2 + \cdots + p_n^2.
\end{aligned}
$$

This polynomial is nonnegative for all x (because it was a sum of squares). Therefore its discriminant $B^2 - 4AC$ must be negative or zero, that is, $B^2 \leq 4AC$, or $(B/2)^2 \leq AC$, which is to say,

$$(p_1 q_1 + \cdots + p_n q_n)^2 \leq (p_1^2 + \cdots + p_n^2) \cdot (q_1^2 + \cdots + q_n^2).$$

How do you like this trick?

OTHER BOOKS IN THE SERIES

Algebra is the third book in this series of books for high school students. The first two, published in 1990, are *Functions and Graphs* and *The Method of Coordinates*. Future books will include:
Pre-Geometry
Geometry
Trigonometry
Calculus

As organized and directed by I. M. Gelfand for a Mathematical School by Correspondence, the books are intended to cover the basics in mathematics. *Functions and Graphs* and *The Method of Coordinates* were written more than 25 years ago for the Mathematical School by Correspondence in the former Soviet Union. Still under the guidance of I. M. Gelfand, the School continues to thrive at such places as Rutgers University, New Brunswick, NJ and Bures-sur-Yvette, France.

As Gelfand himself has stated:
"It was not our intention that all of the students who study from these books or even completed the School by Correspondence should choose mathematics as their future career. Nevertheless, no matter what they would later choose, the results of this mathematical training remain with them. For many, this is a first experience in being able to do something completely independently of a teacher."

Gelfand continues:
"I would like to make one comment here. Some of my American colleagues have been explained to me that American students are not really accustomed to thinking and working hard, and for this reason we must make the material as attractive as possible. Permit me to not completely agree with this opinion. From my long experience with young students all over the world, I know that they are curious and inquisitive and I believe that if they have some clear material presented in a simple form, they will prefer this to all artificial means of attracting their attention—much as one buys books for their content and not for their dazzling jacket designs that engage only for the moment. The most important thing a student can get from the study of mathematics is the attainment of a higher intellectual level."